THE ROBOT

THE LIFE STORY OF A TECHNOLOGY

Lisa Nocks

GREENWOOD TECHNOGRAPHIES

GREENWOOD PRESS
Westport, Connecticut • London

Library of Congress Cataloging-in-Publication Data

Nocks, Lisa.
    The robot : the life story of a technology / Lisa Nocks.
      p.   cm. — (Greenwood technographies, ISSN 1549–7321)
    Includes bibliographical references and index.
    ISBN-10: 0–313–33168–5 (alk. paper)
    ISBN-13: 978–0–313–33168–8
    1. Robots.   2. Artificial intelligence.   I. Title.
  TJ211.N63   2007
  629.8'92—dc22   2006038191

British Library Cataloguing in Publication Data is available.

Library of Congress Catalog Card Number: 2006038191
ISBN-10: 0–313–33168–5
ISBN-13: 978–0–313–33168-8
ISSN: 1549–7321

First published in 2007

Greenwood Press, 88 Post Road West, Westport, CT 06881
An imprint of Greenwood Publishing Group, Inc.
www.greenwood.com

Printed in the United States of America

The paper used in this book complies with the
Permanent Paper Standard issued by the National
Information Standards Organization (Z39.48–1984).

10 9 8 7 6 5 4 3 2 1

Contents

vi   Contents

Series Foreword

In today's world, technology plays an integral role in the daily life of people of all ages. It affects where we live, how we work, how we interact with each other, what we aspire to accomplish. To help students and the general public better understand how technology and society interact, Greenwood has developed *Greenwood Technographies*, a series of short, accessible books that trace the histories of these technologies while documenting *how* these technologies have become so vital to our lives.

Each volume of the *Greenwood Technographies* series tells the biography or "life story" of a particularly important technology. Each "life story" traces the technology, from its "ancestors" (or antecedent technologies), through its early years (either its invention or development) and its rise to prominence, to its final decline, obsolescence, or ubiquity. Just as a good biography combines an analysis of an individual's personal life with a description of the subject's impact on the broader world, each volume in the *Greenwood Technographies* series combines a discussion of technical developments with a description of the technologies effect on the broader fabric of society and culture, and vice versa. The technologies covered in the series run the gamut from those that have been around for centuries—firearms and the printed book, for example—to recent inventions that have rapidly taken over the modern world, such as electronics and the computer.

While the emphasis is on a factual discussion of the development of the technology, these books are also fun to read. The history of technology is full of fascinating tales that both entertain and illuminate. The authors—all experts in their fields—make the life story of technology come alive, while also providing readers with a profound understanding of the relationship of science, technology, and society.

Acknowledgments

No book is the product of one person's labor. I would like to express my gratitude to all of those who shared their expertise, time, or visual materials with me. Special thanks to Ray Bates, MBHI President, The British Clockmaker; Sven Behnke, University of Freiberg Computer Science Institute; Dr. Johann Borenstein, University of Michigan CSE; Dr. Damian Lyons, Department of Computing, Fordham University; Alan Madsen, Shadow Robot Company, Ltd.; Maider Marcos Ortego, Fatronik, San Sebastián, Spain; David Orenstein, Communications & PR Manager, Stanford School of Engineering; and Dave Umberger, Purdue University News Service. Finally, thanks to Kevin Downing, who showed incredible patience through the completion of this project, Nicole Azze and the staff of Greenwood Press, and to Vivek Sood at Techbooks.

Of course, any factual errors are entirely my own.

Introduction

Since the year 2000, major manufacturers including Honda, Toyota, and Sony have exhibited humanoid robots that look like diminutive suited astronauts. Though they can climb stairs, dance, run, communicate with their operators, and even play musical instruments, they are not nearly as sophisticated as the fictional androids depicted in movies like *AI: Artificial Intelligence* (2001), or *I, Robot* (2004). Nevertheless, these entertainment robots have come a long way from the first industrial robotic arms that were integrated into production lines to spray paint and spot weld in the decades after World War II. They surpass the basic criteria for a robot outlined over three decades ago by the Robotics Institute of America: "a reprogrammable, multifunctional manipulator designed to move material, parts, tools, or specialized devices, through variable programmable motions, for the performance of a variety of tasks" (Logsdon, 19). Many researchers have come to see the manufacture of intelligent, "friendly" humanoids that can interact sociably with humans as a viable solution to labor shortages in the caregiving industries, exploration, and warfare. Japanese researchers refer to them not as tools but as "partners" and "friends." Still, humanoid robots that provide anything more than entertainment are a long-range goal, and represent only a very small percentage of the robots in use or development today.

During the last 30 years, Robotics has expanded beyond manufacturing and entertainment to the medical, military, exploration, and service industries. Walking and flying insect-bots, wheeled and track-propelled mobile robots, precision arm and hand robots are only some of the configurations incorporated into such fields as space and underwater exploration, environmental cleanup, mining and excavating, construction and manufacturing, surgical, search and rescue, and combat operations. Robotic mobile tour guides, toys such as Robosapien, robot kits including Lego® Mindstorms and Robonova-1, and robot competitions like Battlebots and Robocup provide entertainment while users supply valuable feedback to developers studying perception, precision, and human-robot social interaction. The newest mobile robots can navigate their environments independently, learning from their experience. In addition, virtual or "cyber" bots (computer software using AI) serve as important instruments for surveillance and information services like Web browsers.

The robots making headlines today are the result of four decades of improvements in robot kinematics, perception, and artificial intelligence. However, the robotic machine is actually a by-product of many centuries of thinking about toolmaking. We have become so accustomed to popular, futuristic representations of robots, that it is hard to believe that the idea emerged in antiquity. In the following chapters, I trace the history of the robot from those early days to the twenty-first-century research labs where robots are learning to cooperate with humans by expressing emotional states.

Part I is an introduction to the forerunners of the robot. Much of the information from the earliest period of toolmaking comes to us as an amalgam of myth and archaeological evidence that demonstrates ancient peoples' interest in technological progress. In Chapter 1, "Between Ritual, Myth, and Machine," I describe the earliest iteration of the automated servant, small funerary statuettes made by the ancient Egyptians called *shabti figures*. These were not machines at all, yet were meant to operate like modern robots, relieving humans of difficult work in the afterlife. By the first millennium B.C.E., the Greeks had incorporated into their mythology stories of sophisticated machines built from forged metals. The people who circulated the first imaginary stories of robot soldiers, housekeepers, and lovers during this period could not have conceived of such machines if they were not already familiar with the simple machines on which they were based. Their experience combining the pulley, wheel, and lever for irrigation, architecture, transportation, and war is reflected first in wondrous tales of robots made by clever artisans and activated by the gods, and later in ingenious real automatic devices or "automata." For instance, almost four centuries after the first telling of the mythical winged flight of Daedalus, a real mathematician investigating the principle of aerodynamics named Archytus of Tarentum (c. 420–411 B.C.E.) built a wooden pigeon automaton powered by a jet stream of compressed air that allegedly could fly.

Although automata demonstrated serious mathematical principles necessary for larger civil projects, they were mainly used to impress the inventor's royal patrons, and for the patrons to amuse their guests. Between the Middle Ages and the Renaissance, the creation of hydraulically powered garden automata led to the production of mechanical ones. During this period, frequent regional wars also inspired schematics and models for land and air vehicles, and more than one model of a robotic knight.

Chapter 2, "From Automata to Automation" begins with the clock and toymakers of the seventeenth through the nineteenth centuries, who responded to the Enlightenment idea of "Man the Machine" by creating figural automata that could write, dance, play instruments, or sing. Though these devices were far less sophisticated than the humanoid entertainment robots of the twenty-first century, their cam-driven programming anticipated modern computing and the first robot controllers. Whereas in slave-based cultures the practical advantages of automatic machines other than for irrigation and astronomical clocks was inconsequential, in the nineteenth century people made the intellectual transition from thinking of automatic machines as merely amusing novelties to recognizing their potential as efficient industrial tools. Automaton makers were some of the first contributors to the mechanization of factories. Their inventions were adapted to steam powered machines to assist and even replace human factory workers in an expanding industrial society. Mechanization was only one aspect of the second industrial revolution. During this period theories of manufacture and labor emerged as movements that turned industry into a science.

The second part of this volume covers early twentieth-century advances in control engineering, early efforts to incorporate robots into industry, and the integration of artificial intelligence into robotics. Chapter 3, "Into the Factory," begins with the 1920s and 1930s, when primitive mechanical men were exhibited by companies like Westinghouse as gimmicks to promote technical innovations in electrical and mechanical engineering and to encourage the public to see technology in a positive light. By the late 1930s, the first robotic arms were patented in anticipation of the need for increased speed and efficiency in wartime manufacturing. Efforts to improve automatic weapons and decipher enemy code resulted in both cybernetics—a theory of control engineering based on biological models—and the binary code that is the basis of the first modern computers.

The invention of transistors and electronic controllers at mid-century made it possible to build truly robotic manipulators. By the early 1960s robotic arms were used in die casting, and to assemble thousands of identical parts for televisions and automobiles in a booming postwar consumer economy. In Chapter 4, "Smarter Machines," I discuss how the concept of artificial intelligence (AI) was established as a subfield of computing, and how competing theories of AI and advances in microprocessing over the next three decades made it possible to adapt vision, audio, and haptic sensors, and controllers to a variety of robotic platforms.

The third and final part of this volume deals with the expansion of robotics to fill a variety of functional niches. Chapter 5, "Getting Around," covers the excitement and challenges of making robots more perceptive and independent. In the 1980s, autonomous mobile cart-type robots equipped with laser range-finders, 3-D stereo camera "eyes" and audio sensors and the latest in AI were introduced into corporate and hospital environments, and a decade later were getting ready to explore other planets. Their value for environmental cleanup would be demonstrated after some nuclear plant emergencies, including the Chernobyl nuclear disaster of 1986.

In response to their exposure to different theories of AI, high-tech computing and engineering equipment, and the need for new technological solutions in different areas of life, a new generation of students adopted the interdisciplinary approach to robotics-introduced decades earlier by the founders of cybernetics. In the late 1980s, an alternative, "bottom-up" approach to mobile robotics emerged. The theory behind what is now called Behavior-Based Robotics (BBR) is that minimally programmed robots that learn from their sensorial experience are more useful and cost-effective. During the last two decades of the twentieth century while BBRs roamed the corridors of research labs and eventually the surface of Mars, researchers began working in earnest on an idea once relegated to fiction—the humanoid robot. A number of labs experimented with approaches to bipedalism (two-legged motion) and dexterous hands as the most useful robotic tools. Others concentrated on perceptual systems, synthetic muscles, or different approaches to AI and robot learning. In the early 1990s grant-based research began at MIT on the first BBR humanoid, COG, while Japanese engineers pursued humanoid projects begun earlier and funded by manufacturing companies and the Japanese government.

After an "AI Winter" during which trust in the goals of artificial intelligence withered, the invention of neural net chips and further reductions in the size and price of microchips allowed for a new wave of innovation in robotics. By this time there were humanoid robot projects in progress in at least ten countries, funded in part by military subsidies, and in part by R&D capital from heavy machinery companies. At the end of 30 years humanoid robots had learned to do something that it took us millions of years to do—stand up and walk on two feet. Chapter 5 concludes with a history of this infant subfield of robotics, and locates it within the growing field of autonomous mobile robots.

Since the beginning of the third millennium, military agency representatives, health care providers, and business people have participated in government-sponsored hearings on the future uses for robots. Robot developers regularly participate in international trade fairs where they demonstrate prototypes for a variety of robotic platforms and configurations. Innovative engineers have formed companies to manufacture robotic pool cleaners, vacuums, surgical instruments and prosthetics, remote-presence "house sitters," animatronics for movies and theme parks, and toys. Chapter 6, "The Latest Developments" features the work of the researchers who are slowly bringing these new robot configurations to maturity using state of the art sensors, artificial muscles, and learning theories borrowed from the behavioral and cognitive sciences. Here in the beginning of the twenty-first century they have achieved in material terms the idea expressed in the Homeric legend, *The Iliad*—robots equipped with "intelligence, speech, and strength" that can "do a great many things." I also describe the ways that we are becoming socialized to accept a robotic culture, from the rising popularity of robotic toys to robot developers' manipulation of language in their advertisements. I conclude with a brief reflection on the achievements and failures in the field during the last quarter of a century, and speculations by researchers about the future of robots.

Space limitations prevent more than a survey of the growth of the robot. For example, I have not included the rather complex story of labor relations surrounding the introduction of robotics into industry, nor the institutional history of robotics and AI labs. I encourage the reader to investigate these subjects, as well as the works consulted to prepare this volume. I have also included a few Web sites that will both inform and inspire further interest in this most fascinating and timely subject. Some of these sites provide access to videos of robots in action. A comprehensive timeline of robotics and a glossary of relevant technical terms are also included.

Timeline

c. 4000–3000 B.C.E.	The first simple machines come into use in Mesopotamia.
c. 3500 B.C.E.	Water-raising lever systems in use in Babylonia, Assyria, Egypt, and India; wheels for pottery and transportation in use in Mesopotamia, India, and China.
c.2500 B.C.E.	The wheel in use in Egypt.
c. 2100 B.C.E.	Egyptian Middle Kingdom—Shabti figures introduced.
c. 1500–1400 B.C.E.	Inclined planes used as ramps in Egypt at the great temple of Hatshepsut.
c. 1500–1000 B.C.E.	The pulley in use.
c. 800–700 B.C.E.	Homeric oral epic *The Iliad* written down; mentions robot-like machines.
c. 700–600 B.C.E.	Automatic theaters with water effects are built in Nineveh, Assyria.
c. 422 B.C.E.	The clepsydra (water clock) is used for the first time.
c. 400 B.C.E.	Archytus constructs a pulley-driven wooden dove automaton powered by compressed air.
384–322 B.C.E.	In *Mechanical Problems* Aristotle describes multiple-wheel pulleys and cogwheels.

332 B.C.E.	Alexander the Great conquers Egypt where he founds the city of Alexandria.
300 B.C.E.	Ptolemy I Soter founds the Alexandrian School.
300–c. 200 B.C.E.	First Alexandrian period.
	Euclid systematizes the theorems of plane and solid geometry.
	Archimedes of Syracuse, Sicily describes the *worm gear*, calculates pi (π), articulates the mathematical principle of the lever, and develops the principles of hydrostatics.
	Ctesibius improves the clepsydra and produces automata powered by compressed air.
250–180 B.C.E.	Philo of Byzantium writes on mechanics.
230 B.C.E.	Multiple cogwheels in use in China.
c. 200 B.C.E.– c. 200 C.E.	Second Alexandrian Period.
1st century C.E.	Hero/Heron of Alexandria produces *On Automatic Theatres, Mechanics, and Pneumatics*; produces a kind of steam engine; Antikythera mechanism (astronomical calculator) produced (by Germinus of Rhodes?)
c. 476–c. 1350	Middle Ages (from the fall of the Roman Empire to the Renaissance).
c. 800	Abdullah al-Ma'mūn, Caliph of Baghdad, commissions *Kitab al-Hiyal* (*The Book of Ingenious Devices*).
13th century	Ramon Lull (1232–1316) attempts to formalize thought via mathematics.
1206	*Treatise of al-Jazarī* of Persia on automata.
1350–1650	The Renaissance (between the Middle Ages and the Modern era).
1300s	Jacquemarts incorporated into European tower clocks.
1497	Leonardo draws a schematic for a robot knight.
1515	Leonardo demonstrates a lion automaton to King Francis I of France
c.1540	Giannello Torriano builds a knight automaton for the exiled King Charles V.

1600s	A mechanical organ first described by Kircher.
1600–1867	Japanese Edo Period; hundreds of automata are built, including tea carrying dolls.
1642	At 19, Blaise Pascal designs an adding and subtracting calculator and sells 50 of them.
1672	Gottfried Wilhelm Leibniz designs a calculator that adds, subtracts, multiplies, divides, and finds square roots.

THE EIGHTEENTH CENTURY

1719	A silk mill in operation in England is considered the first modern factory.
1726	In *Gulliver's Travels*, Jonathan Swift describes an automatic book-writing machine.
1738	Vaucanson demonstrates three automata in Paris.
Mid-1700s	The Enlightenment; also known as the Age of Reason.
1750s	Friedrich von Knauss produces the first known writing automata.
1764	James Hargreaves invents a spinning jenny that spins several threads at once.
1770s	Pierre Jaquet-Droz produces a writing android with a self-contained mechanism.
	Henri Maillardet produces writing and fortune-telling androids.
1790	The partnership of Matthew Boulton and James Watt results in an improved version of Thomas Newcomen's steam engine.
1795	Tippu's Tiger automaton is discovered in India.
1796	In Japan, Yorinao Hosokawa publishes a three-volume manual, *Sketches of Automata*.

THE NINETEENTH CENTURY

1805	Joseph Marie Jacquard invents punch card programming for automated weaving looms.

1825	Building on the work of Faraday, Oersted, Arago and others, William Sturgeon invents an electromagnet as a teaching aid. This leads to the invention of the electric motor.
1828–1839	Charles Babbage designs his Difference Engine; then with Lady Lovelace begins an Analytical Engine to calculate astronomical tables. Neither is completed.
1830	American Christopher Spencer designs a cam-operated lathe.
1854	Mathematician George Boole publishes *The Laws of Thought*.
1868	Zadoc P. Dederick and Isaac Grass of Newark, New Jersey receive a patent for a steam-driven cart pulled by a mechanical man; it is meant for public transportation. A few months later, the first story about a "steam man" is published in *Beadles Dime Novels*.
1871	Charles Babbage dies, leaving over 400 sq ft of drawings for his Analytical Engine.
1877	Lord Kelvin (William Thomson) demonstrates that machines can be programmed to solve mathematical problems.
1892	Seward Babbitt designs a motorized crane with gripper to remove ingots from a furnace.
1893	Canadian George Moore builds a walking steam man meant to pull heavy loads.
1898	Nikola Tesla demonstrates his model for a remote-controlled submarine at Madison Square Garden in New York City.

THE TWENTIETH CENTURY

1929	First humanoid exhibition robots shown in England and the United States.
1937	Alonzo Church and Alan Turing independently develop a thesis that is widely interpreted as evidence for equivalence between human and machine intelligence.
1938	Willard V. Pollard files a patent for "Position Controlling Apparatus," a robotic arm.
1939	The patent for Harold A. Roselund's automatic spray-painting arm is granted to the DeVilbiss Company.

Exhibition robot ELEKTRO and dog SPARKO featured at the New York World's Fair.

1940   John Atanasoff and Clifford Berry design ABC, the first all-electronic, nonprogrammable computer; British computer research group Ultra builds ROBINSON, an electromechanical relay computer used to crack the German code.

1941   In Germany, Konrad Zuse develops the Z-3, the first operational programmable digital computer, programmed by Arnold Fast.

1942   Isaac Asimov describes the Three Laws of Robotics in a short story, "Runaround."

1943   Warren McCullough and Walter Pitts produce a mathematical model of neural net computing in their "Logical Calculus of the Ideas Immanent in Nervous Activity."

Ultra builds COLOSSUS, an electronic tube computer many hundreds of times faster than ROBINSON, and which can decipher more complex German codes.

Researchers gather at Princeton to discuss biologically based feedback control systems.

William Grey Walter and his wife Vivian begin work on mobile robots that can display spontaneity, autonomy, and self-regulation to understand animal nervous systems.

1944   MARK I by Howard Aiken, the first programmable computer built by an American, uses paper tape for programming and vacuum tubes for calculations.

1945   Professor John von Neumann of Princeton University publishes a paper describing the concept of the stored program computer.

1946   ENIAC, the world's first fully electric, general-purpose reprogrammable digital computer is developed for the Army by John P. Eckert and John W. Mauchly. It is almost 1000 times faster than the MARK I.

George Devol patents a general-purpose playback device for controlling machines that uses a magnetic process recorder.

The first Macy conference on biologically based machine control is held in New York.

1947   The transistor is invented at Bell labs by William Bradford Schockley, Walter H. Brattain, and John Bardeen, setting off the microelectronics revolution. It can switch currents on and off at far greater speeds than vacuum tubes.

U.S. National Labs in Argonne develop the first teleoperated manipulator (Master/ Slave Manipulator).

1948   Norbert Wiener publishes *Cybernetics*, a book on information theory that he defines as "the science of control and communication in animal and machine."

Grey and Vivian Walter exhibit their first mobile robots, which they call "tortoises" or "Machina speculatrix," to the press.

1949   An essay on Cybernetics by E. L. Locke appears in *Astounding Science Fiction*.

EDSAC, by Maurice Wilkes at Cambridge is the world's first stored program computer.

BINAC is developed by Eckert and Mauchly's new computing company.

1950   UNIVAC by Eckert and Mauchly is the first commercially marketed programmable computer. A programmable computer compiles the U.S. census for the first time.

In "Computing Machinery and Intelligence," Alan Turing describes a method for determining machine intelligence that will come to be known as the "Turing Test."

1951   EDVAC, built by Eckert and Mauchly at the Moore School, U Penn., is the first stored program computer built in the United States.

A Cybernetics Congress is held in Paris, France.

In France, Raymond Goertz designs the first teleoperated mechanical arm for the Atomic Energy Commission—a milestone in force feedback technology.

1953   First marine remotely operated vehicle (ROV) is used for underwater photography.

1954   George Devol develops the first reprogrammable robotic arm.

Planet Corporation is formed to manufacture robots.

1955	Planet Corp. installs a PLANETBOT, the first polar coordinate commercial robot arm, in GM's Harrison radiator manufacturing plant.
1956	John McCarthy, Marvin Minsky, and others organize the first interdisciplinary conference on AI at Dartmouth.
1958	Jack St. Clair Kilby creates an integrated circuit at Texas Instruments.
	McCarthy develops an AI program, LISP.
	Seymour Cray builds the CDC 1604, the first fully transistorized supercomputer, at the Control Data Corporation.
	The United States establishes the Defense Advanced Research Projects Agency (DARPA) a major source of funding for both computing and robotics research.
1958–1959	Jack Kilby and Robert Noyce independently invent the computer chip.
1959	John McCarthy and Marvin Minsky found the MIT AI Lab.
	Computer-assisted manufacturing is demonstrated at the Servomechanisms Lab at MIT.
1960s–1970s	Robotic marine ROVs are in development.
1960	George Devol's Unimation is purchased by Condec Corporation and begins development of UNIMATE robot systems.
	Theodore Harold Maiman constructs the first laser. It uses a ruby cylinder.
	DARPA makes significant increases in funding for computer research.
	American Machine and Foundry (AMF) Corporation markets the first cylindrical robot, VERSATRAN, designed by Harry Johnson and Veljko Milenkovic.
1961	Joseph Engelberger founds Unimation Company, Inc. in Danbury, CT. UNIMATE, the robotic arm invented by George Devol and marketed by Unimation, is installed for die casting in a GM plant in Ternstedt, Ewing Township—a suburb of Trenton, New Jersey.

1963	John McCarthy founds the Stanford University AI Lab.
	Primitive neural nets, using only a few synthetic neurons, show limited capability.
	The RANCHO ARM, the first computer-controlled prosthetic arm is invented. Its six joints give it human-like flexibility.
	Digital Equipment Corp. (DEC) develops the PDP-8, the first successful minicomputer.
1964	Gordon Moore correctly predicts integrated circuits will double in complexity annually.
1965	Herbert Simon predicts that in 20 years machines will be capable of doing any work a human can do.
1966	The SHAKEY mobile robot research project (1966–1972) begins at Stanford Research Institute under the directorship of Charles A. Rosen.
	June 2, the first SURVEYOR mobile robot makes a soft landing on the moon.
1967	The Swedish company, Svenska Metallverken (ABSM) is the first European company to purchase a UNIMATE.
1968	Marvin Minsky develops a tentacle robot arm at MIT.
	Kawasaki Industries, established as a ship builder in 1896, begins production of hydraulic robots under a license agreement (1968–1985) with Unimation.
1969	At Stanford Research Institute Victor Scheinman develops the first electrically powered computer-controlled robot, referred to as the "STANFORD ARM."
	Sweden imports its first UNIMATE.
1970	Serbian mechanical engineer Miomir Vukobratović proposes a theoretical model, *Zero Moment Point,* to explain and control biped robot locomotion.
1971	Intel introduces its 4004, the first microprocessor.
1973	THE TOMORROW TOOL (T3), the first commercially available minicomputer-controlled industrial robot, is developed by Richard Hohn for Cincinnati Milacron Corporation.

The AI department at Edinburgh, United Kingdom, demonstrates their robot FREDDY II assembling objects automatically from a heap of parts.

At Waseda University, Japan Ichiro Kato unveils the first humanoid, WABOT-1.

1974   "SILVER ARM," which uses feedback from touch and pressure sensors, is designed for small-parts assembly.

Victor Scheinman forms Vicarm, Inc. to market a minicomputer-controlled version of the Stanford Arm to industry.

The Swedish company ASEA (later part of ABB Automation Technologies) launches the IRB-6, the world's first microcomputer-controlled, all-electric industrial robot.

The Robotics Industries Association (RIA), the only trade group in North America organized specifically to serve the robotics industry, is founded in Ann Arbor, Michigan.

1975   Automatix Company is founded to produce industrial robots.

MYCIN, a medical expert system for the diagnosis of infectious blood diseases is developed by doctoral student Edward Shortliffe.

1976   Vicarm Inc. incorporates a microcomputer into its robot arm.

Robot arms on Viking 1 and 2 Mars Landers take soil samples.

Ray Kurzweil produces the first text to speech reading machine for the blind using optical character recognition (OCR) software.

Stephen Wozniak and Steven Jobs found Apple Computer Corporation.

1977   Unimation purchases Vicarm, Inc.

ASEA markets two electrically powered, microcomputer-controlled industrial robots.

1978   With support from General Motors, Unimation develops the Programmable Universal Machine for Assembly (PUMA) based on Vicarm technology.

	Texas Instruments releases SPEAK & SPELL, a learning device for young children. It is the first commercial product that electronically duplicates the human vocal tract on a computer chip.

1979   Hans Moravec creates the Stanford CART, an autonomous vehicle that can navigate across a room by detecting and avoiding obstacles.

MYCIN is said to work as well as human medical experts by *Journal of the American Medical Association.*

1980s   Honda Motor Company begins its first humanoid P-series robot research program.

Expert systems, software programs that can access and analyze specialist knowledge, are in wide use.

The neural network paradigm is revived; researchers are using multiple layer networks.

1980   At Case Western Reserve University, Roger Quinn and Roy Ritzmann develop "ROBOT III," a six-legged robotic insect.

1980–1982   The first MIT Leg Lab robot, the Planar One-leg Hopper with a small foot and controlled with a simple three-part algorithm, is designed to explore active balance and dynamic stability in legged locomotion.

1981   IBM introduces its Personal Computer (PC).

1983   Together, 30 robot manufacturers in Germany produce 2000 industrial robots.

1983–1984   MIT Leg Lab builds a 3-D one-leg hopper.

1984   A musician humanoid prototype WABOT-2 is demonstrated at Waseda University.

Stanford AI lab develops FLAKEY, a second-generation mobile robot with real-time stereovision algorithms to distinguish and follow people, and the DECIPHER speech recognition system to respond to spoken commands.

As part of its Strategic Computing Initiative, DARPA launches its Autonomous Land Vehicles (ALV) initiative to develop robotic vehicles for use in war. During the same period, Bundeswehr University develops an ALV project funded by German automobile and electronics companies.

At Microelectronics and Computer Technology Corporation (MCC), Douglas Lenat begins inputting common sense knowledge into a computer to help robots understand our world.

The Apple MACINTOSH desktop computer system is released.

1985–1990   MIT Leg Lab develops a simple biped robot to study locomotion on rough terrain, running at high speed, and gymnastic maneuvers.

WABOT-2 performs every day at the Science Exposition in Tsukuba, Japan.

1986   Initially funded under the DARPA ALV project, Carnegie Mellon University begins its NavLab research, producing many prototypes over two decades.

Japanese Industrial Robotics Association (JIRA) estimates for the end of 1986 (excluding fixed-sequence and manual manipulators) are Japan: 116,000, the United States: 25,000, Germany: 12,400, France: 5,270, and Sweden: 2,380.

1987   The Shadow Company, United Kingdom, begins its humanoid project, SHADOW BIPED.

Robotic vision systems are now a $300 million industry.

1988   Stäubli Group purchases Unimation from Westinghouse.

Rodney Brooks, director of the MIT AI lab, and his student, Colin Angle, begin work on GENGHIS, an insect-like six-legged walking robot.

Computer memory costs only one hundred millionth of what it did in 1950.

Danny Hillis' CONNECTION MACHINE, a massively parallel processor, can perform 65,536 computations simultaneously.

1989   Rodney Brooks and Anita Flynn's paper, "Fast, Cheap and Out of Control" inspires others to look at simpler yet efficient robot designs.

At JPL, Colin Angle, Rajiv Desai, and David Miller construct the wheeled planetary rover prototype TOOTH.

1990s

1990   ROBODOC robotic arm by orthopedic surgeon William
            Bargar and researcher Howard Paul is used for a hip
            replacement surgery on a dog.

            Robotic vision systems are now an $800 million industry.

1992   ROBODOC successfully used in a human hip replacement
            and is approved by the FDA.

            Howard Paul forms Integrated Surgical Systems to
            commercialize ROBODOC.

            The Humanoid Project, a joint venture of Japanese
            academia, government, and industry evolves from the
            Waseda University humanoid lab.

1993   At MIT, Rodney Brooks begins COG, a humanoid robot
            that will learn like a child.

            The new Intel Pentium 32-bit microchip has
            3.1 million transistors.

1994   DANTE II, a six-legged walking robot developed at
            Carnegie Mellon University Robotics Institute explores the
            Mt. Spurr volcano in Alaska and samples volcanic gases.

            Doug Lenat, now consulting professor at Stanford
            University and president of Cycorp, Inc. continues the
            project he began in 1984 as CYC (for
            encyclopedia).

            DEC introduces a processor that can execute one billion
            instructions per second (BIPS).

1995   Fred Moll, Rob Younge, and John Freud form Intuitive
            Surgical to design and market surgical robotic systems using
            technology based on the work at SRI, MIT, and IBM.

Mid-1990s   Hydro-Quebec Research Institute is developing compound
            camera-equipped teleoperated submersible robots to inspect
            underwater parts of dam walls.

            Neural computing chips are on the market.

1995   Waseda University, Japan researchers develop HADALY-1
            humanoid prototype to share space with humans.

1996   Honda's humanoid Prototype 2 (P-2) can walk, climb stairs,
            and carry loads.

In December, NASA launches Mars Pathfinder with SOJOURNER robotic rover.

1997   In January, NEC Corp. begins R100 remote presence personal robot (now PA PE RO).

In May, world chess champion Garry Kasparov loses to IBM's DEEP BLUE supercomputer.

Waseda researchers produce HADALY-2 humanoid and WABIAN, which uses improved vision sensors to recognize several gestures.

The first Robocup robot soccer match is held in Nagoya, Japan.

On July 4, SOJOURNER, the 22-pound microrover is dispatched from Pathfinder on the surface of Mars, and sends back images of its travels.

Honda debuts its eighth humanoid prototype, P3.

NASA space agency begins developing ROBONAUT, a legless humanoid meant to assist astronauts outside space vehicles and space stations.

NASA successfully uses AERCam Sprint free-flying robot camera in the cargo hold of the STS-87 space-walk.

1998   Virtual robots (computer screen personalities) capable of responding to spoken and visual cues from human users are in development for service and information industries.

MIT doctoral student Cynthia Breazeal begins her PhD project, KISMET, a face robot modeled on a human infant using vision and speech to study human–robot socialization.

Toshiba unveils its half-ounce, camera-equipped microrobot for retrieving unwanted particles from passages as small as 25 millimeters.

Tiger Electronics introduces FURBY, the first interactive "smart" pet.

Waseda researchers develop WENDY humanoid prototype, an advance on HADALY.

Microvision tests its VIRTUAL RETINA DISPLAY, which is meant to project images directly onto the user's retinas.

1999   Sony sells out the initial run of its $2500 AIBO robotic dog.

THE TWENTY-FIRST CENTURY

2000   NASA Ames Research Center near San Francisco is developing a free-flying personal assistant robotic sphere for use on the International Space Station.

Triton XL marine ROV is used to prevent seepage from the sinking oil tanker Erika.

The United Nations estimates that there are 742,500 industrial robots in use worldwide.

Honda debuts its humanoid, ASIMO, and Sony introduces its humanoid, Sony Dream Robot (SDR) and a second-generation AIBO pet robot at the first Robodex trade show.

Many other robotic pets are introduced including dogs iCYBIE by Silverlit of Hong Kong, Tomy's DOG.COM, WEB WEB by Lansey (United States), Sega's POO-CHI; and cats Bandei's BN-1, Tiger's ANIM, and Omron's NeCoRo.

Robotic baby doll MY REAL BABY, a joint venture of Rodney Brooks' iRobot Company and toy manufacturer Hasbro goes on the market along with MGA Entertainment Corporation's MY DREAM BABY and Mattel's MIRACLE MOVES BABY.

2001   The Space Station Remote Manipulator System (SSRMS) by MD Robotics of Canada is successfully launched to complete assembly of International Space Station (ISS).

In April, the GLOBAL HAWK robotic spy plane charts its own course over a distance of 13,000 km (8,000 miles) between California, United States, and Southern Australia.

Canadarms 1 and 2, which can lift objects in space at weights equivalent to a space shuttle, shake hands during an ISS mission.

Fujitsu debuts Humanoid for Open Architecture Program (HOAP-1), an approximately 18-inch tall humanoid to be sold to researchers for around $41,000 each.

Kawada Industries, Japan, introduces the HRP-2P an approximately 5 foot tall LINUX-based open-architecture humanoid meant for work rather than entertainment.

In New York, Professor Jacques Marescaux performs the world's first telesurgery (a gallbladder removal) using a ZEUS controller on a patient in Strasburg, France.

Search and rescue robots used at Ground Zero after the September 11 attacks on the World Trade Center include marsupial bots by Professor Robin Murphy (University of South Florida); shape-changing recognizance robots (a Colorado team led by Colonel (retired) John Blitch); and reconfigured mini pipe inspection robots by Inuktun (British Columbia).

2002   The UNECE estimates 350,000 industrial robots in use in Japan, 233,000 in Europe, and 104,000 in the United States.

iRobot's ROOMBA, a disc-shaped robotic vacuum, goes on the market.

A Japanese research consortium demonstrates a humanoid robot driving a backhoe.

Waseda researchers are working to instill emotional cues and responses in humanoid prototypes WAMOEBA and WABIAN.

Sony demonstrates SDR-4X android at Robodex 2002.

2003   April 7 is the fictional birth date of the Japanese cartoon robot MIGHTY ATOM, known to Westerners as ASTRO BOY. Japanese researchers used the date as a goal for releasing the first successful humanoid robots.

BHR-1, a 5-foot tall martial arts entertainment robot is created by Li Kejie and team at the Beijing University of Science and Engineering.

WAKAMARU, Kawasaki Heavy Industries home companion robot is exhibited at ROBODEX 2003, along with humanoid prototypes from Honda, Sony, and others.

2004   UN World Robotics Reports two-thirds of the domestic robots in use were purchased in 2003; and that by the end of 2007, 4.1 million domestic robots will be in use, about half of these will be entertainment robots.

FANUC, the largest Japanese robot manufacturer, estimates it has sold 120,869 robots.

CENTIBOTS, 100 autonomous mobile robots by SRI, successfully work as a team to map, track, and guard in a coherent fashion over a 24-hour period.

Robot rovers SPIRIT (January 3) and OPPORTUNITY (January 24) land on Mars.

ROBOSAPIEN, an interactive, programmable humanoid by veteran NASA engineer Mark Tilden goes on the market.

Brian Carlisle and Bruce Shimano form Precision Robotics to produce a new generation of sensors, controllers, and high-precision robots to handle very small parts.

2005   Shadow robot begins shipping its robotic DEXTROUS HAND.

March 25–September 25, Toyota Partner Robot, Sony QRIO, and Honda ASIMO perform together on stage for the first time at the EXPO 2005 in AICHI, Japan. Visitors to the information booth are greeted by a female android, ACTROID.

The Robotics Industries Association (RIA) estimates the United States is now second only to Japan in industrial robot use with 158,000 robots installed in U.S. manufacturing operations.

Toyota unveils its "PARTNER" robot initiative by demonstrating humanoid musicians.

A Stanford University team led by Sebastian Thrun wins the DARPA Grand Challenge with their ALV STANLEY.

Tokyo-based ZMP, Inc. releases the small home companion robot, NUVO for $6000; an army of NUVOs appear in an episode of the TV program, *Crossing Jordan*.

WAKAMARU, a LINUX-powered home companion humanoid from Mitsubishi Heavy Industries goes on sale in Japan for 1.5 million yen (approx. U.S. $14,000).

2006   In January, Sony announces it will discontinue AIBO manufacture and QRIO development in order to redirect capital to other consumer lines.

Part I

ANCESTORS

1

Between Ritual, Myth, and Machine: Robot Ancestors

◆

If every instrument could accomplish its own work, obeying or anticipating the will of others . . . if the shuttle could weave, and the pick touch the lyre, without a hand to guide them; chief workmen would not need servants.
—Aristotle (384–322 B.C.E)

In 1921, the Czech writer Karel Čapek first presented *Rossum's Universal Robots* (R.U.R.), his play about humanoid workers that are manufactured by the thousands. Čapek's robots were organic beings, not mechanical men. Nevertheless, they served the same purpose as the real robots introduced into factories just a few decades later: to relieve human beings of difficult, monotonous, or dangerous work. After the play was shown internationally, the term "robot" (roughly "worker" or "wage slave" in Czech) was adopted by social critics to refer to human beings who do mindless, repetitive work. It also became popular among science fiction writers as a way to describe the intelligent automatic machines that were appearing more frequently in their stories. By the 1960s, engineers began to apply the term "robot" to re-programmable industrial machines that could do a variety of repetitive tasks independent of an operator. Recent advances in robotics have narrowed the line between science fiction and technological fact.

We have become so accustomed to thinking of robots as part of our future that it may be difficult to believe that they were conceived over

4,000 years ago. Like all sophisticated technologies, robot engineering is the result of many centuries of accumulated knowledge about the laws of physics, mathematics, and geometry. Robots are also a creative response to the problem of labor. The first incarnation of the robot was not a machine in our sense of the term; but it was a clear expression of the idea of manufacturing machines to replace humans in laborious tasks.

RITUAL

Over thousands of years, civilizations had grown up around agriculture; and in the absence of sophisticated machinery, all the individuals in a community had to share the responsibility for irrigating fields and planting and harvesting crops. Since Egyptians envisioned the afterlife as a transcendent likeness of the earthly one, they expected that this responsibility would continue for eternity. As in most cultures, the wealthy and powerful had servants to work for them, and according to Egyptian religion, this practice continued in the netherworld. Before the First Dynasty (−2686 B.C.E.), human servants were entombed with their dead masters for this purpose. (Some Egyptologists believe that the servants were put to death at the time of the king's burial.) Gradually this practice was replaced by painting images of servants on tomb walls. Later, the dead were buried with statuettes depicting servants engaged in different tasks. Finally, the *shabti*, a stylized, mummiform figurine, was introduced during the Middle Kingdom (c. 2100 B.C.E.).

At first only one or two high quality, custom-made shabti, usually of stone, were included in each tomb. These were considered a mark of high status. Despite the fact that the popularity of the shabti had spread across class lines, the level of artistry gradually declined. However, during the eighteenth Dynasty of the New Kingdom (c. 1550 B.C.E.), kings adopted the practice, and the production of high-quality shabti resumed. By this time, shabti made of stone, faience (glazed earthenware), or painted wood were being produced in the thousands. They were equipped with hoes, grain baskets, and other necessary tools, although over time, these accessories were simply carved or painted onto the shabti. The figurines originally ranged in height from just a few centimeters to around 50 centimeters (19.7 inches); however, as demand increased, the production of shabti was standardized at a smaller size. In later years, so many of them were being entombed with each body that additional "overseer" shabti were included in the ritual, presumably to supervise the others.

The shabti initially served both as the deceased's servant and as an extra home for the "ka" (an ambulatory "reserve" body that obtained nourishment for the deceased from tomb offerings). By the New Kingdom, they were understood to be solely servants for the deceased. Surviving sales records for shabti from the Third Intermediate Period (c. 1069–650 B.C.E.) refer to them as "male and female slaves" that once paid for would work in the afterlife for the named deceased. Egyptians believed the shabti would perform what we would describe today as robotic action: automatically rising when called to work, and returning to rest afterward. This is reflected in the transformation of the term from shabti to *ushebti* ("answerer" or "respondent") during the later years of the New Kingdom. Their "program code" was a magical incantation prescribed in the *Book of the Dead*, "The Chapter of Not Doing Work in Khert-Neter":

> Illumine the Osiris Ani, whose word is truth. Hail, SHABTI FIGURE! If the Osiris ani be decreed to do any of the work which is to be done in the Khert-Neter, let everything which standeth in the way be removed from him—whether it be to plow the fields, or to fill the channels with water, or to carry sand from [the East to the West]. The SHABTI FIGURE replieth: I will do it, verily I am here [when] thou callest. (*Book of the Dead*, XIII, 629)

During the New Kingdom, the "Speech of the Ushabti Figure" was frequently inscribed on the figurines. The idea of the manufactured worker continued to surface in the area around the Mediterranean. By around 1000 B.C.E., the Greeks had developed complex oral traditions that included tales of intelligent, autonomous machines made to serve the gods.

MYTH

Scholars of ancient myths interpret the adventures of gods and demigods, witches, and demons as metaphorical explanations of the origin and workings of the natural world. Despite their imaginative and symbolic nature, we should not disassociate myths from the history of technology. Within their symbolic structure, we can derive clues about people's attitude toward toolmaking. For instance, in the Homeric epic *The Iliad* (written down c. 800–700 B.C.E.) the deformed Hephaestos, god of the forge, produces both female attendants of gold "in appearance like living women" with intelligence, speech, and strength, and wheel-driven tripod stools that move automatically to serve the gods at table. The depiction of Hephaestos' robots as servants is consistent with the Greeks' slave-based culture. Furthermore, unlike the ushebti that required craftsmanship but not mechanical expertise, Greek mythology acknowledged the importance of technical ability as requisite to the success of civilization. In another tale, Hephaestos creates the bronze statue Talos, which Zeus gives to his consort Europa. She takes it to the island of Crete, which is ruled by King Minos. Talos, activated by oil in a vein that runs from his heel to his neck, serves as a robotic sentry, making rounds of the island three times a day. When Talos encounters Sardinian warriors attempting to invade the island, he grows red-hot and destroys them by forcing them into a burning embrace.

The idea that Talos is given as a gift to a person of power (in some versions directly to King Minos) reflects the way that bronze objects were reserved for the elite classes by the time *The Iliad* was first told. Talos' physical stature and power may imply the importance of practical metallurgy for civil and military applications. Scholars think that the invaders who attacked Greece from the north around 1200 B.C.E. used weapons of iron rather than bronze. Although we cannot know to what extent these tales were inspired by historical events, it is possible that the earlier transition from bronze to iron tools is implied in the story of Talos' demise: as he tries to fend off the attack of the Argonauts who have iron weapons, the witch Medea causes the nail stopper that holds in Talos' life-giving oil to fail. It drains from his ankle, rendering his bronze body useless.

Within a few hundred years of the Homeric legend, the Greeks began to produce complex automatic machines, the forerunners of robots. They could not have conceived of building these "automata" without a working knowledge of what we call the "classical" or "simple" machines.

MACHINES

In the ancient world, as people formed communities around agriculture and trade, they recognized the need for more effective tools. For instance, the forging of bronze (an alloy made from copper and tin) and then iron made possible the production of stronger, sharper tools for domestic and military purposes. These had a great advantage over earlier stone implements. Other activities like raising water for irrigation, drinking, milling grain, or moving stone to erect buildings—all required the great effort of many people. The number of people may not have been an issue in slave-based civilizations, but efficiency was. This inspired the development of the first machines, which came into use in Mesopotamia during the fourth millennium B.C.E., and spread at varying rates over the next two thousand years to civilizations in Asia, Africa, and Europe. Thus Marcus Vitruvius Pollio (c. 70–31 B.C.E.), a Roman engineer and architect writing in the first century B.C.E., could state with accuracy that machines were "well known." Vitruvius defined a machine as "a connected construction of wood, which gives very great advantages in raising heavy masses . . ." A machine does not do away with the need to exert effort, but by helping to distribute weight or create leverage, it makes the effort more productive than the exclusive use of muscle power. It therefore increases the amount of useful work that can be accomplished over a given period of time. Vitruvius enumerates the "simple machines" as the lever, the inclined plane, the wedge, the pulley, and the toothed wheel. (The basic wheel is a simple machine; the addition of teeth made it more versatile.) Since the principles used in these machines are incorporated into modern technology including robots, it is useful to understand how they work.

The Greek mathematician Archimedes (287–212 B.C.E.) is often credited with articulating the mathematical principle of the lever. The *lever*, which had long been used for lifting boulders, is a rigid board or bar that rests on a fulcrum (turning point). Pivoting the bar at the fulcrum and applying effort at one end shifts the load at the other end. For instance, by pivoting an automobile jack handle against the jack (fulcrum), the opposing bar lifts an automobile (load) off the ground. The principle of the lever was also applied for water raising. In this case, a weight is attached to the shorter arm of a two-armed lever, and a bucket to the longer arm. The arms are set over a tall vertical beam (fulcrum). The workers apply effort to the bucket to dip it in the water. Releasing the full bucket causes the weight at the other end to drop, lifting the bucket of water from its source. The lever can be rotated to dispense the water to its target. This arrangement, called a *shaduf* among the Babylonians, Assyrians, and Egyptians, and a *picotah* among the Indians, was used as early as 3500 B.C.E. A lever system for raising water is depicted on a Babylonian cylinder seal of the Akkadian period (c. 2400–2200 B.C.E.), and on the surviving palace walls of Nineveh of the seventh century B.C.E. The same system was still in use in Egypt in the twentieth century. In the first century B.C.E., Philo of Byzantium (born c. 250 B.C.E.) described a treadle version in which a rocking platform is attached to the shorter end of the lever. The long arm with the bucket is dropped into the water, and as a person walks up the treadle creating a counterweight, the bucket rises. In a more sophisticated, double pole system, the person walks up one treadle and then moves down the other, allowing one bucket to dip while the other, full bucket, rises.

The lever creates and maintains equilibrium (balance). For instance, a person can carry a large amount of water more effectively by distributing the weight equally between two buckets placed on either side of a beam. The beam hangs an equal distance on each side of the person's shoulders. This idea of seeking equilibrium led to the use of the lever system for weighing objects. A doctor's scale with its stationary and sliding weight system is a modern form of the earliest weight machine, called a *steelyard*. In the ancient device, the object to be weighed was suspended from the shorter lever arm, which turns on the fulcrum. A counterweight slides along the longer arm that is marked in increments. When the counterweight reaches equilibrium, the notch where it stops indicates the weight of the object. These are all first-order levers, in which the fulcrum sits between the load and the effort. A wheelbarrow is a second-order lever. In this case, the load is placed between the wheel (fulcrum) and the handlebar where the effort is exerted. In a third-order lever, the effort is applied between the fulcrum and the load. An example of a third-order lever is the application of our muscle (effort) to lift our forearm (load) at the elbow (fulcrum). Robotics engineers have done much work in simulating the third-order lever in robotic arms with artificial muscles, discussed later in this volume.

The *inclined plane* is a smooth surface with one end set higher than the other. This may not seem like a machine because it makes no apparent movement. Yet it does perform work by literally lifting the burden, making it possible for people or animals to move heavy objects from one level to another without carrying them up or down steps or lifting weight vertically over long distances. Archaeologists theorize that during the Old Kingdom, the Egyptians used inclined planes as ramps to drag stones up to the lower levels of the pyramids. They were certainly in use in Egypt by the fifteenth century B.C.E., for the remains of the great temple of Hatshepsut indicate a solid ramp rose to its third level. This shape is also useful for other work: When it tapers off to a sharp edge, it can be used as a *wedge* as in a crowbar. Once a wedge is placed at the end of a long handle and pivoted on a fulcrum, it becomes a lever. Early people placed primitive crowbars under stones to raise them more easily. The chisel, hatchet, and axe were in use since primeval times for dividing objects into pieces, for instance in chopping wood.

The *screw* is an inclined plane that winds around itself in a helix or spiral, also called a worm. Carpentry screws, which have many winds (threads), are used to hold things together. In the ancient world, this shape had another useful function. The importance of water for drinking, washing, milling, and irrigating fields made it necessary to invent ways of getting it from its source to other locations. The hydraulic (water) screw was originally adapted to this purpose: A very large screw that is open at both ends is encased in a watertight cylinder. One end of the screw is placed in the water, and the other end elevated at a certain angle. As it turns, the water trapped in air pockets between the screw threads rises and is released through the upper open end of the device. We can be reasonably certain that this machine was invented during the first millennium B.C.E., although scholars disagree whether it originated with the Assyrians (c. 800–600 B.C.E.), Archytus of Tarentum (c. 420–350 B.C.E.), or the mathematician and physicist, Archimedes of Syracuse (287–212 B.C.E.).

The *wheel* is one of the most versatile of the simple machines. Drawings on pottery and pictographs on tomb walls indicate that the wheel was being used with an axle to make pottery and for transportation in Mesopotamia, India, and China by 3500 B.C.E. and in Egypt after 2500 B.C.E. A pictograph of a wheeled cart appears in the account records of the Inanna Temple in Erech (Sumer) Mesopotamia from this period, and the remains of a four-wheeled hearse (sledge on wheels) dating from c. 3000–2000 B.C.E. were found with the skeleton of a draught animal in a tomb in what was Kish, Mesopotamia. A statuette of a wheeled oxcart from c. 2500–2000 B.C.E. was unearthed at Mohenjo-daro, once a major city in the Indus Valley. The wheel was in use in the Aegean world around the time of the Homeric legends. The axle developed from the roller, probably originally adapted from tree trunks. Rollers were used alone or under a sledge (sled) for rolling heavy objects from place to place; and would have been most useful in conjunction with the inclined plane. People discovered that objects could be moved more easily by running a thinner roller through the center of two wheels, creating an axle. Turning the wheels on an axle allows for more stability and smoother change in direction. Over time, the wheel was transformed from a solid, heavy disc to a lightweight though more resilient spoked form like those used on chariots for military and other civil uses.

The *pulley* combines the rope with the wheel to lift heavy objects. Rope has been used since the Paleolithic period for hunting and fishing, binding, and moving objects. People discovered the advantage of hoisting a heavy object on a rope by pulling down on it rather than by pulling the object up. Prior to the invention of the wheel, the rope was probably simply drawn over some smooth surface like a log for leverage. Later it was discovered that winding a rope around a grooved wheel mounted on a block creates further mechanical advantage. The groove or channel keeps the rope in place as it moves. The pulley is represented in Assyrian art of the eighth century B.C.E. However, archaeologists think it was in use before the first millennium B.C.E., since the modern Arabic word for pulley is found on a fifteenth-century B.C.E. tablet from Alalakh. People eventually discovered that the weight distributed across additional wheels creates even greater advantage. Aristotle described a multiwheel pulley in the fourth century B.C.E. Another iteration of the pulley is the *capstan*, a vertical spool-shaped revolving cylinder for hoisting weights by winding in a cable. It was used both aboard ships for drawing up anchors, and for milling, where it was driven by animals or slaves.

Vitruvius, who devoted several chapters of his *de Architectura* (first century B.C.E.) to an explanation of how the simple machines work, acknowledged their importance to the efficient operation of a society:

> Without the aid of wheels and axles, of presses and levers, we could enjoy neither the comforts of good oil, nor the fruit of the vine. Without the aid of carts and wagons on land, ships on the sea, we should be unable to transport any of our commodities. How necessary also, is the use of scales and weights in our dealings, to protect us from fraud. Not less so are innumerable different machines, which it is unnecessary here to discuss, since they are so well known from our daily use of them . . . (Vitruvius, Book X, Ch. I, 5.)

COMPOUND MACHINES

Around the first millennium B.C.E. people were using compound machines to advance the work of raising and pouring water, calculating distance and time, moving heavy building materials, and fighting battles. For instance, the block and tackle, a combination of ropes and pulleys, were rigged like cranes to lift extremely heavy stones. Levers and pulleys were combined to make catapults, the first artillery weapons. The *cogwheel*, a simple wheel with pegs or teeth attached around its circumference, is actually a combination of the lever and the wheel. When a cogwheel is set in motion against another one, its teeth act as levers, pulling back on the cogs of the next wheel, forcing it to change direction. Depending on the relative size of the wheels, the speed as well the motion of the second wheel will change. Two or more interlocking cogwheels working together form a *gear*, which increases the number of useful actions a basic wheel can perform. We know cogwheels were already in use in the fourth century B.C.E. since Aristotle (384–322 B.C.E.) described them in his *Mechanical Problems*. There is evidence that multiple cogwheels were used in China from c. 230 B.C.E. Other gear forms were also in use during this period. For instance, during the third century B.C.E., the mathematician Archimedes described the *worm gear*, a spiral shaft (screw) combined with a cogwheel; and Philo of Byzantium described a primitive chain and sprocket drive (like those used on modern bicycles) for repeated loading and discharging of a catapult.

The *watermill* was one of the first semiautomatic compound machines. It derives its power from the *waterwheel*, which looks like an amusement park Ferris wheel: a vertical wheel with paddles or buckets turns at the surface of the water, scooping it up as the wheel turns. The weight of the water dropping onto the paddles or buckets keeps the wheel turning. The power of the vertical wheel is translated via a gear to horizontal millstones, which turn against each other to grind raw grain. The challenge was the regulation of the prime mover (water). The original waterwheels were "undershot," meaning that the water flowed from its source under the wheel. Once the wheel was turning, the weight of the water falling over the back of the wheel kept it turning into the water to continue the process. However, at certain times of the year the water level dropped, interfering with the wheel operation. One solution was the "overshot" system, in which water was regulated by damming its flow further upstream, and directing it via a shoot over the top of the wheel. A much later solution was the "floating" waterwheel, which accommodated the seasonal changes by moving up and down with the water level.

A few centuries after the Homeric poet first related Daedalus' attempted winged escape from Crete, Archytus of Tarentum (born c. 420–411 B.C.E.) used his theory of pulleys to build a wooden dove automaton. The bird allegedly could flap its wings and fly, powered by a jet stream of compressed air. Unlike Hephaestus' mythical robots, "ingenious devices" like this were not animated by gods, but by human beings. Between the fourth century B.C.E. and the rise of the Roman Empire in 30 B.C.E., mathematicians, physicists, and craftsmen from the areas around the Mediterranean applied the principles of physics and geometry to produce a variety of automatic machines. While they served to amuse their patrons, these devices were also material demonstrations of the practical value of hydraulic (water), pneumatic (compressed air), aerostatic (air/wind), and mechanical action. The same principles were later applied to machines like catapults, irrigation apparatuses, steam engines, and clocks. Much of this early work in mechanics was a product of the Alexandrian school.

THE ALEXANDRIAN SCHOOL

During the final centuries of the first millennium B.C.E., Greek culture was centered in cities outside of Greece, in Northeastern Africa and Western Asia, and especially Alexandria, Egypt. Alexander the Great, son and successor to King Philip II of Macedon and pupil of Aristotle, was enamored of Egypt. After conquering it in 332 B.C.E., he founded a city there, which he named after himself. Following Alexander's untimely death in 323 B.C.E. one of his generals, Ptolemy I Soter (c. 367–283 B.C.E.), secured Egypt for himself. There in c. 300 B.C.E. he founded the Alexandrian School, a research complex that included a museum (shrine to the Muses, the gods of intellect) and a library of almost half a million manuscripts. Until a series of fires destroyed the library, the complex was the center of Hellenistic culture and the foremost center of learning in the ancient world. Here, during the first Alexandrian period (c. 300–200 B.C.E.) for instance, Euclid (c. 300 B.C.E.) systematized the theorems of plane and solid geometry.

The mathematician Archimedes (287–212 B.C.E.) of Syracuse, Sicily, spent his youth in Alexandria, where he wrote a number of mathematical treatises, among them a work on hydrostatics. This "theory of displacement" states that the force acting to buoy up a body immersed in water is equal to the weight of the water displaced. He also is credited with defining π (pi), adapting the principle of the lever to catapults, and inventing the hydraulic screw described above. Archimedes invented a hydraulic organ, a planetarium, and a series of connected devices meant to drag a large, heavy ship ashore. The first industrial robotic arms recall another of his inventions, the *manus ferrea* ("iron hand"). This machine was equipped with a claw connected to a crane that was meant to raise and overturn enemy ships.

Ctesibius (born c. 270–250 B.C.E.), an Alexandrian barber who took up the study of mechanics, hydraulics, and pneumatics, made key contributions to the history of automated machines. Using these principles, he invented a pneumatic organ and the first water clock to incorporate moving figures. Descriptions of Ctesibius' works survive in the pneumatics section of *Mechanical Collection* (c. 200 B.C.E.), written by his contemporary and disciple, Philo of Byzantium. Vitruvius also describes Ctesibius' inventions, among them a water pump for automatic fountains. Ctesibius' own manuscripts must have still been available at the Alexandrian library in the first century B.C.E. because in his description of the water pump, Vitruvius notes that those interested in finding out about Ctesibius' entertainment automata can read the inventor's works for themselves. Among the most interesting of Ctesibius's constructions was a pneumatically driven singing bird based on the principle that he had discovered during his career as a barber. Ctesibius had invented an adjustable mirror for his shop that used a counterweight suspended inside a tube. He noticed that the contraption made a whistling noise as the air escaped when the weight moved through the tube. This inspired more experiments with air pressure, and resulted in the first known singing bird automaton, later included in automatic fountains.

One of the most famous of Ctesibius' inventions was his *clepsydra* ("captured water") or water clock. The clepsydra, first used around 422 B.C.E., was a basic siphon initially used for experiments in physics, and as an egg timer. During the first Alexandrian period, it was adapted as a way for physicians to count the pulse. It was also used in law courts to time speeches: A long tube was plunged into the water and when it was full, the opening at the top was closed. When it was reopened, the water dripped through a small opening at the lower end. A person was free to speak until the tube was empty. Theoretically, the interval between drips marked a specified time; however, the rate of flow increased when there was more water in the tube. As it emptied, the decrease in pressure slowed the dripping. Ctesibius' objective was to regulate the clock so that the water level did not have to be continually tended. He used a three-tier system in which a large body of water emptied into the clepsydra to insure it remained full. A float and pointer set in a third container indicated the time elapsed. Ctesibius' clepsydra remained the most accurate clock until the fourteenth century when mechanical clocks using a system of leaded weights and levers replaced hydraulic ones. The float in the clepsydra represents an early example of a *feedback* mechanism.

Heron (or Hero) of Alexandria has been called the greatest Greek engineer of the Second Alexandrian Period (until c. 200 C.E.). Heron taught at the Museum, where he promoted the idea of combining theoretical and practical knowledge. He built on the works of Ctesibius, Archimedes, Euclid, and others of the earlier period; and is credited with a number of inventions including a float regulator to control the opening of temple doors and the movement of automatic theatres. These are the product of his conviction that mathematics is the basis for practical engineering.

Heron produced a number of manuscripts including *The Automaton Theatre*, *Mechanics*, and *Pneumatics*. In *The Automaton Theatre*, he described a puppet theater controlled by strings, drums, and weights. In the two-volume *Pneumatics*, he described a number of automatic devices that worked by air, steam, or water pressure. Among them was "A Bird Made to Whistle by Flowing Water," perhaps a hydraulic version of Ctesibius' compressed air singing bird. Another was an automatic holy water dispenser. When a coin of a certain weight (a 5-drachma piece) was dropped in a slot, it activated a lever system that automatically opened a lid of the vessel, allowing water to flow out through its spout. Once the coin slid to a certain point, it would trip another lever, closing the lid, and stopping the flow of water. Heron even described at least one automaton that ran by solar power, "A Fountain which Trickles by the Action of the Sun's Rays" (Heron, *Pneumatica*, 47).

During the same period, astronomers were making progress with geared devices that calculated distance and time. As ancient people ventured beyond their shores for trading and military expeditions, the astronomical charts used to project information about the weather and tides for farming were adapted to navigation. The ability to chart the stars from any position provided sailors with a method of finding their location and keeping on course. One of the earliest and most impressive astronomical calculators for which we have material evidence is the *Antikythera mechanism*, a first-century invention named after the small island between Crete and the Greek mainland where in 1900 the device was recovered from the wreck of a commercial ship. It is the only known configuration for an astronomical calculator built before c. 1000 c.e.

The device measures about 32 by 16 by 10 centimeters (12.6 by 6.3 by 3.9 inches) and comprises 32 brass gears mounted in a wooden box. Although it was crushed in the wreck, the deposits that formed around the eroding metal gears preserved their shape, helping researchers to deduce the original mechanism. A differential gear provided their relative position at each phase of the moon, so that when the operator turned a crank to input a future or past date, the machine calculated the position of the sun, moon, and other astronomical information. It is likely that the device was being used as a navigational tool aboard the ship that sank. Some researchers date it to within a decade of the disaster (c. 87 b.c.e.) and link it to Germinus of Rhodes through similarities between inscriptions on the device and those in one of his surviving manuscripts. In its heyday, Rhodes was known as a center for astronomical thought. It was the home of Hipparchus (c. 170–125 b.c.e.), considered the father of systematic astronomy, and Posidonius (c. 135–50 b.c.e.), who determined the nature of the tides and allegedly built an astronomical computer even more complex than the Antikythera mechanism. The skill of Rhodian astronomers was so well established that it is incorporated into the Homeric epics. It is not surprising that this first "computer" was a navigational tool. As we will see in upcoming chapters, improvements in astronomical tables and calculating machines aided the British navy in expanding its military and mercantile advantage in the nineteenth century.

MEDIEVAL AUTOMATA

The achievements of all of these ancient inventors were introduced into Europe and the Indian subcontinent during the medieval period, when Arab scholars were collecting and translating Greek manuscripts; and testing and elaborating on their work. For instance, in the early ninth century Abdullah al-Ma'mūn (786–833 c.e.), the reigning Caliph of Baghdad, commissioned the *Kitab al-Hiyal* (*The Book of Ingenious Devices*). This description of over one hundred devices was compiled by three men—the Banū (sons of) Mūsa—from Greek texts preserved in monasteries and in areas around Mesopotamia and brought to Baghdad.

The sons of a noted astronomer, the Banū Mūsa, were educated under the patronage of the Caliph after their father's death and are acknowledged as key contributors to Arabic science and technology, and astronomy. They produced at least 20 of their own manuscripts dealing with mathematics, geometry, astronomy, the construction of related measuring devices, and a description of a musical automaton. After acquiring some wealth, they sent missions to Byzantium to bring back manuscripts of ancient writers, and allegedly paid a regular salary to translators who worked in the House of Wisdom where they themselves had studied. Among their most important translations are ancient Greek texts on automatic machines, including Hero's *Mechanics*. Their writings indicate that they were also familiar with the work of Philo and Archimedes. *The Book of Ingenious Devices*, attributed mainly to the brother Ahmad, the engineer of the family, is a series of annotated schematics for different types of liquid dispensers. Ahmad's book was well known in medieval Islam, and was often referred to as an important work.

Continued interest in automatic hydraulic and pneumatic devices is apparent in the much-admired *Treatise of al-Jazarī* (1206 C.E.), which was copied several times over the centuries into Turkish and Persian. By his own account, al-Jazarī worked in the court of the Sultans of Āmida from the year 1181 C.E. to 1182 C.E., but it is not clear whether all the devices he describes were invented during that period for particular sultans, or whether some of the designs were copied from Greek inventors Archimedes and Apollonius, or from the Banū Mūsa. Al-Jazarī describes a variety of automatic, water-powered devices that incorporate levers and pulleys including water clocks, cups for washing and bloodletting, perpetual flutes, fountains, and instruments to raise water from a flowing stream. An elaborate peacock-shaped vessel for washing the hands includes a mechanism that first causes a miniature servant to appear with soap, then another with a towel. Another amusing automaton represents two male figures sitting in a throne room drinking. It is a kind of water-wheel device in which the cups of the seated figures are filled, lifted, and emptied in a circuit about eight times per hour.

One of the most ingenious automatons was designed to serve wine to a king at special events. It consists of a figure of a girl in a building that resembles a guardhouse. A copper dome above the house is a chamber meant to be filled with wine. A tube acts as conduit for the wine, which drips into a reversible cup. Eight times per hour the cup fills, tips over into another container from which another tube passes the wine into a glass drinking-cup held by the figure. When set in motion by a series of weights, levers, and pulleys, the figure opens double doors, rolls down a ramp (inclined plane), and serves a cup of wine. In its opposite hand is a handkerchief for the king to wipe his mouth. Once the king drains the cup and uses the handkerchief, he sets them back in place and pushes the figure back up the ramp. The rolling motion causes the girl's right hand to rise and fall, and eventually to engage a hook that keeps the figure in place against the inner wall of the house while the cup refills. The action can be repeated until the wine chamber is empty. This perhaps seems trivial, until we consider that each of the movements is important to modern machinery. For example, the automaton's movements are regulated (as in a clockwork) to operate according to specified time increments. Moreover, al-Jazarī notes the importance of the hinge joint of the girl's arm, which allows for the dual action of raising and lowering the arm and extending it to the king. This demonstrates that inventors were already working with the problem of degrees of freedom eventually accomplished in modern robotic manipulators.

By the time monks began copying Arab manuscripts in the monasteries of Europe, medieval chroniclers were remarking on the existence of unique automata. For instance, in the thirteenth century, William of Malmesbury related how before Gerbert of Aurillac became Pope Sylvester II (999–1003), he and a servant had discovered an underground stash buried centuries earlier. Among the objects were golden knights that played dice and defended themselves when approached by the explorers. We cannot know whether William of Malmesbury was relating a legend, or embellishing on earlier descriptions of actual automata. However, we do know that during this time, both Christian and Hebrew scholars were trying to establish the correct place of human mechanical and creative activity, and that people were telling stories about the creation of human figures that could move of their own accord.

RENAISSANCE AUTOMATA

During the Renaissance (1350–1650), Latin translations of the Alexandrian and Arab mechanics inspired the work of a great many European inventors; among them was Leonardo da Vinci (1452–1519). According to his chronicler Lomazzo (1584), a student and heir of Leonardo named Francesco Melzi remarked on flying bird automata made by Leonardo, possibly inspired by his reading translations of Archytus of Tarentum. According to Lomazzo, Leonardo also built a lion automaton that walked a distance, stood up, made a noise, and opened its chest to reveal a bouquet of lilies, (the "fleurs de lis" of France). The lion was allegedly demonstrated to King Francis I of France in 1515 in honor of an alliance between his kingdom and Florence.

Leonardo's *Codex Atlanticus* includes a design for a mechanical knight clad in German–Italian medieval armor. The schematic indicates articulated joints and moveable head, jaw, and arms, and indicates a pulley system for providing the robot's appendages degrees of freedom. Next to this is a sketch of a human leg with a pulley wheel at the knee joint, an indication that Leonardo adapted the human joint to the proposed robot. Any explanation Leonardo may have given for powering the robot has not survived, and there is no evidence that he ever actually built it. However, he did inspire others. Around 1540, Giannello Torriano (1515–1585) built a knight automaton for the exiled Charles V that incorporated a system of cables and pulleys. More recently, Mark Rosheim, an engineer who has done robotics work for NASA, recreated Leonardo's knight from the schematic, first as a computer simulation and then as a model. His interest in Leonardo's work inspired the robot that he has been developing for use on the International Space Station.

Leonardo also made sketches and possibly working models for a number of other automatic machines, including a glider, a bird, and a leaf-spring powered cart that may have served as the platform for the lion. The cart seems reminiscent of the self-moving tripods built by Hephaestos in the Greek legend discussed above. Leonardo seems to have also been inspired by the Greek legends to produce schematics for his flying machine. He would have known the legend of Daedalus who constructs a set of wings and escapes while imprisoned on Crete after provoking the ire of King Minos.

The translation and distribution of Greek and Arabic texts into Latin also inspired the Italian commentaries on automata published in the sixteenth and seventeenth centuries. For instance, an automatic mechanical organ described in the 1600s by Kircher "which utters the voices of animals and birds" is an updated version of the hydraulic theaters and organs of Archimedes and Heron. Mechanical devices that moved, uttered sounds, and played music were popular among the elites for entertaining their guests. They were frequently made both for the dinner table and for estate gardens. Craftsmen combined hydraulic, pneumatic, and mechanical principles to produce fountains with singing birds, and mechanical people and animals that moved around the grounds. During the seventeenth century the philosopher Descartes (1596–1650) described one of the most impressive automatic parks, designed by the Florentine engineer Thomas Francini for the Château Saint-Germain-en-Laye, west of Paris. Here hydraulically powered figures drawn from mythology moved and even squirted water when visitors walking through the garden tripped a mechanism embedded in the tiled walkway. Hellbrunn Park in Salzburg, Austria, built in the seventeenth century by the Archbishop Markus Sittikus von Hohenems, contains a number of "trick" elements. The designer of its automata is unknown, although they may possibly have been the work of the fountain master of the estate project, the monk Fra Gioachino. Hellbrunn is still in working order and is a popular tourist attraction.

The idea of the robot was reflected in medieval and Renaissance fiction. People made of metal are featured in mid-twelfth-century French fiction such as *Le conte de Floire et Blancheflor* and Benoît de Sainte-Maure's *Roman de Troie* (c. 1165). Possibly influenced by the Homeric legend of Talos, many of the fictional humanoids were enlisted as guards. For instance, in the early thirteenth-century work *Lancelot do lac*, the hero spars with two copper knights and confronts a copper woman in his quest to overcome the enchantment on the fortress, Doloreuse Garde. In Rabelais's (1490–1530) *Gargantua and Pantagruel*, Gargantua spends some of his time contriving "a thousand little automatory engines, that is to say, moving of themselves" (Rabelais, Book I, Ch. XXIV). A particularly amusing robot story from the end of the period was written by Giambattista Basile (1575–1632), a writer best known for *Lu Cunta de li Cunti* (*The Tale of Tales*, 1634–1636). This collection of stories is often referred to as *Il Pentamerone* after Boccaccio's *Decameron* with which it shares a literary framework. Among the 50 tales, supposedly being told to a prince and his bride over a five-day period is the story of Bertha. In a twist in the Pygmalion myth, the maiden Bertha continually ignores her father's wishes that she marry, until one day when she asks him to bring her large quantities of sugar, ambrosia almonds, scented water, musk, ambergris, gold thread, pearls, rubies, and some other precious gems. From these Bertha creates a handsome young man. She entreats the Goddess of Love to bring the man to life, and her request is granted.

Basile's tale appeared around the same time as some clockwork automata which are thought to have been produced in Spain. One of them is a lady that appears to walk gracefully while turning her head and strumming a cittern (similar to a mandolin). The control mechanism is housed beneath her skirt, and hidden notched wheels drive her feet, giving the appearance of walking. The maker of this work is unknown, but it has sometimes been credited to Giannello Torriano, who was inspired by Leonardo's knight automaton, and who allegedly built a number of automata to entertain Charles V. A second surviving piece of unclear origin is a walking monk. Like the cittern player, it is governed by a clockwork mechanism, and hidden wheels drive its feet. It moves in a 2-foot square pattern, nodding its head, and moving its eyes and mouth. It beats its chest with one hand while raising and lowering a crucifix with the other.

All of the ingenious devices described here anticipated the fascinating and expansive body of automata produced during the eighteenth and nineteenth centuries. Like their antecedents, they had no apparent purpose other than amusement. Yet the mechanical principles upon which they were based are derived from improvements in clocks. The clock and automaton makers, respected for their mechanical ability, made many contributions to mechanization of factories.

2

Automata to Automation

◆

It may sometimes happen that the greatest efforts of ingenuity have been exerted in trifles, yet the same principles and expedients may be applied to more valuable purposes...

—Dr. Samuel Johnson (1709–1784)

The ingenious devices discussed in Chapter 1 amused inventors' wealthy patrons; but more importantly, they demonstrated practical engineering solutions. Indeed, in the Middle Ages, invention had become "a total and coherent project ..." for large numbers of technicians who had begun to "... consider systematically all the imaginable ways of solving a problem" (White, 173). Many had come to view the universe as a bottomless reservoir of natural energies that once identified, could be tapped for practical purposes. It was during this time, for instance, that Guido da Vigevano sketched a war chariot driven by wind power. Over the centuries the idea of wind-powered vehicles fell by the wayside; but wind power itself would be adapted to other areas, such as agriculture. In fact, the search for reliable power sources is one of two key problems that have continued to challenge engineers. The second was to invent regulators to make those machines more efficient, that is, to produce more work with less effort. For instance, the surviving notebooks of architect and engineer Villard de Honnecourt (flourished c. 1225–1250) contain schematics for a number of mechanical devices including an automatic water-powered saw, a weight-driven clock, and the unicorn of the technological imagination: a *perpetual motion machine*.

When Honnecourt drew his schematic, he acknowledged that the concept of a machine that could run by itself for an infinite period had already been studied for centuries. Among his predecessors was the Hindu astronomer and mathematician Bhaskara, whose influence was a cultural belief in the endless and cyclical nature of life. In 1159, he described two perpetual motion wheels driven by a combination of the two highly fluid elements, mercury and water. Bhaskara's model inspired Arab treatises on similar devices, and it is probably through those works that medieval Europeans were encouraged that the problem of perpetual motion could be solved and adapted to practical activities. However, as the scientist Peter of Maricourt noted in his treatise on magnetism (1269), many had tried and failed to build such a machine. Maricourt himself had devised two of them: one based on magnetic force, and another using a globular lodestone. These designs were an early expression of a modern technical goal: to move beyond mechanization, which requires the constant intervention of human beings, to automation, which requires little or no operator control once a machine is turned on. Although perpetual motion machines are now considered chimerical, they illustrate the dedication of inventors to finding sources of power and regulation that would eventually lead to the invention of the steam engine and electricity. In the Middle Ages, investigations into automation and regulation resulted in improvements in timekeeping.

CLOCKS

The first-century astronomical calculators like the Antikythera mechanism described in Chapter 1 provided the mechanical basis for timekeepers based on gears rather than hydraulics. Information about them slowly spread to other civilizations. The first astronomical tower clock for which we have any detailed information is an important example of the transition from hydraulic to mechanical clocks. It was built in the eleventh century in China by Su Sung and his collaborators.Encased in the tower were the clockwork, a rotating armillary sphere, a celestial globe, and puppets that announced hours and quarter hours with visual and auditory signals. All these mechanical elements were driven by a large waterwheel, also encased in the tower. This hybrid astronomical mechanism, possibly the most complex one ever built, lasted into the fourteenth century. It is unclear whether the device was destroyed or simply deteriorated from disuse. Though it had survived the political takeover by the Chin and the Yuan, the rulers of the Ming dynasty had no interest in automata, and did not keep it in working order. By that time, however, mechanical clocks were being produced in Europe.

The greatest challenge facing clockmakers had been the regulation of a driving mechanism that could operate for long periods of time. The sundial for example had no moving parts but was useless at night or during winter months especially in northern countries. As for hydraulic clocks, their incremental dripping slowed as the water supply was exhausted; and the water would freeze during the cold European winters. Perhaps inspired by designs for perpetual motion machines, a number of late thirteenth-century European clockmakers attempted to solve this problem by using mercury in place of water. Around the same time, the first mechanical clocks were invented. These falling weight clocks also had inherent limitations. Not only did the heavy weights have to be raised by hand at the beginning of each cycle; but also the force of gravity caused them to accelerate as they dropped. Over the following centuries a variety of *escapements* or regulating mechanisms were introduced to overcome these problems.

In the earliest escape system, the driving weight is suspended from a rope that is wound around the axle of a gear wheel (called the *crown* or *escape* wheel). The speed at which the weight drops is regulated by two interact-ing elements: The *verge* is a vertical shaft suspended by a cord or rope and equipped with a pair of paddles or *pallets* set at about 100° angles to each other and spaced relative to the diameter of the escape wheel. Its counter-part, the *foliot*, is a crossbar equipped with adjustable balance weights at each end and set at the top of the verge. The distance the balance weights are set from the center of the crossbar determines the period of *oscillation*, or movement to and fro. The escape wheel drops freely for about 2° until one of its teeth makes contact with a pallet on the verge. As the tooth slides across the pallet exerting force on it, the foliot begins to rotate. It continues to rotate as the lower pallet comes into contact with a lower gear tooth. This pallet pushes the escape wheel in the opposite direction briefly, thus checking the acceleration of the drive weight. Simultaneously, the escape wheel continues to exert force on the pallet. The foliot stops, changes di-rection, accelerates, and rotates about 100°; and the pallet allows a tooth to escape. This process, which keeps the clock "ticking" in equal increments, continues until the weight has dropped completely.

Despite the escapement, these clocks lost many minutes a day. Two serious problems that affected precision were weather and microorganisms. Since the lubricating oils used during this period were not processed, they developed bacteria that contributed to the corrosion of the clock parts. Friction increased when the corroded elements came in contact with each

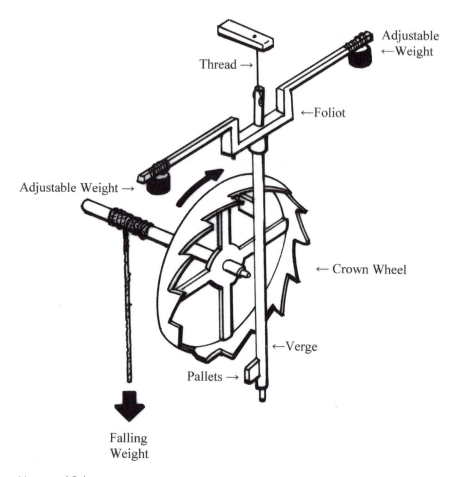

Thread →

Adjustable
←Weight

←Foliot

Adjustable Weight →

← Crown Wheel

←Verge

Pallets →

Falling
Weight

Verge and Foliot

other, reducing the accuracy of the movement. Furthermore, in warm weather, thermal expansion of the wrought iron caused the clocks to lose time; and in cold weather, the opposite effect caused them to gain time. Although these escapements were still in use until the early 1600s, they were gradually replaced with a metal ring called a *balance wheel*. This escapement was also affected by changes in the weather: The crossbar expanded in warmer temperatures, distorting the ring into an oval. However, though the expanded part of the ring caused the clock to lose time, the narrower part caused the clock to gain time. Thus one distortion compensated for the other, reducing the discrepancy in time.

A few innovations provided for more reliable escapements. The problem of friction was reduced with parts made of brass, a less corrosive alloy of copper and zinc. The weight was replaced with a flexible steel wire or *mainspring*, invented in last decade of the fifteenth century by the Nürnberg locksmith Peter Henlein (1480–1542). This made it possible to produce smaller clocks, but it suffered the opposite problem as the falling weight mechanism: as the spring unwound, it lost tension, slowing the mechanism. The introduction of the *stack freed*, and later the more successful *fusee*, invented c. 1525–1540 in Prague by the Swiss mechanic Jacob Zech, exerted a compensating pressure on the spring as it unwound. The pendulum, used first in the sixteenth century by Leonardo da Vinci and his contemporaries to create a reciprocating motion in pumping machines, was first applied to clocks during the seventeenth century. In 1641, Galileo's son Vincenzio built a pendulum clock based on his father's ideas. A competing version of the pendulum mechanism was produced in 1657 by the Dutch physicist and astronomer Christian Huygens (1629–1695) who articulated the relationship between mass and force and eventually invented the more accurate compound pendulum. The pendulum added the force of gravity to the recoil in the earlier escapement, providing a more reliable alternative to the verge and foliot. The presence of two restoring forces reduced the negative impact of friction between the escape wheel and pallet.

Around 1675 a modification of the verge was invented. This *anchor* escapement reduced the angle of swing of the pendulum from 100° to about 6°, requiring less energy to keep it moving. In addition, the pallets were set further away from the axis of rotation, thus requiring a lower angle of rotation to make the same arc and cutting down on the amount of friction throughout the mechanism. Around the same time, Huygens invented a hairspring and balance wheel mechanism that made the production of portable clocks and watches practicable.

Like the Antikythera and hydraulic clocks, the first mechanical clocks were built to keep track of the movements of the planets and stars as an aid in navigation, religious ritual, or agriculture. One of the most famous of these, the 24-hour astrarium (astronomical clock) built in Padua by Giovanni di Dondi (c. 1348–1364), included a perpetual calendar to anticipate moveable religious feasts. Like Su Sung, di Dondi meticulously described how to build and maintain the device.

CARILLON

Early mechanical tower clocks were designed to automatically ring on the hour, loud enough to be heard at a distance. In monasteries, monks were called to prayer at designated times throughout the day and evening by clocks fitted with sets of tuned bells called carillon. A contemporary chronicler noted in 1355 that the 24-hour clock of the church of Saint Gothard in Milan struck a clapper on the hour day and night. Contemporaries of Giovanni di Dondi's father, Jacopo, remark that the mechanical clock he designed for the Palazzo Capitano in Padua (c. 1345) chimed throughout the day and night. By the second half of the fourteenth century, church clocks across Italy chimed at equal intervals over each 24-hour period. Gradually, the striking of the hour became important for the regulation of civil and commercial activities. King Charles V had a chiming clock installed in the Royal Palace in Paris (1370), and in two other locations; and insisted that church clock chimes be coordinated with those of civil clock towers.

By the mid-fourteenth century, European mechanics were enhancing their clocks with moving figures. The Europeans were inspired by an earlier practice incorporated into religious instruction. During the medieval period, moving figurines were sometimes installed near the lectern in churches as a kind of multimedia presentation. The priest could dramatize his sermon by working the figure via a series of weights and levers. In one case, a carved wood crucified Christ rolled its head and eyes and stuck out its tongue to illustrate the agonizing death he had endured for humanity's redemption.

Mechanized clock figures were less instructional than they were outward displays of the complexity and ingenuity of the unseen inner clock mechanisms. One popular figure was the crowing cockerel, a reference to the tradition that Jesus admonished the apostle Peter that he would deny knowing his master "three times before the cock crows." For instance, the astronomical clock of the Strasbourg Cathedral in France (c. 1354) was enhanced with a painted ironwork cockerel that crowed on the hour via a hidden bellows system; and the city clock tower in Berne, Switzerland displayed a crowing cockerel (installed c. 1530). The Berne cockerel was only part of a complex of moving figures that included a parade of bears in varying action poses set on a revolving turntable that moved with the passing of the hours.

JACQUEMARTS

Many of the European cathedral tower clocks displayed humanoid figures called *jacquemarts* ("Jacks of the Clock"). These "Jacks," which had one or two articulated (movable) limbs, were fastened to pedestals or revolving platforms connected to the clockwork mechanism. Some were always visible while others were hidden and appeared from inside the clock at preset times.

When it was time for the clock to chime, one or more of the figurines would draw an arm against some object like a forge or bell as if the figure was producing the sound.

Jacquemarts frequently represented soldiers, a reference to the living ones that guarded the city from the watchtowers. For example, "Jack the Smiter" (c. 1480), set in the tower of Southwald Church in Suffolk, England, is dressed in armor similar to that worn by soldiers during the Wars of the Roses.

This Jack was designed to strike a bell with a hammer to sound the hour. A set of impressive Jacks also graced Old St. Paul's Cathedral, London. Although they were destroyed or stolen during the Great Fire of 1666, their existence is confirmed by references to them in literature of the period, including Shakespeare's *Richard II*. Among the most impressive jacquemarts of the sixteenth century were those made by goldsmith and clockmaker Hans Schlottheim (1547–1625) in Augsburg, Germany; and by the brothers Isaac and Josias Habrecht, mathematicians and watchmakers from Schaffhouse, Switzerland, who built the Salzburg clock (1574). During the following centuries, a number of impressive musical astronomical clocks enhanced with jacquemarts were produced, including those by English technicians Jacob Lovelace (1656–1716) and Henry Bridges, famous for his *Microcosm* (c. 1730–1741).

Miniature moving figures were also used to embellish tabletop clocks. For example the 25-inch high "Passion Clock" built in Augsburg (c. 1580) features tiny figurines of Mary Magdalene, St., John, the Virgin Mary, and four Roman legionnaires that circle around the clock on the hour, gesturing as they pass beneath a model of the crucifixion mounted atop the clock. Schlottheim is credited with a miniature hexagonal tower clock of gilded copper (unfortunately destroyed in the bombing of Dresden in 1945). The nativity characters were motionless until the striking of the twelfth hour, when Joseph would begin to rock the cradle, the shepherds and magi would move forward and bow to the Christ child, and Mary would make a gesture of welcome. A globe above the clock would then open, revealing an image of God bestowing a benediction; while elsewhere angels ascended and descended from heaven, and a musical instrument played a lullaby of its own accord. An equally impressive device by Schlottheim was a *nef* built around 1580 for Emperor Rudolph II. Originally, nef were vessels shaped like ships that were meant to hold utensils or wine on a dining table. Over time, they were enhanced with clocks and moving figures, transforming them into grand conversation pieces. Schlottheim's nef, approximately 2.5 feet long, contains a gilt brass clock enhanced with several miniature figures that move in a procession around the clock face and before a throne.

"Jack the Smiter" jacquemart, Southwald Church Tower, England c. 1480. © Ray Bates. Used by permission.

The clockmakers of the Black Forest region of Germany had a tradition of decorating tabletop clocks with moving figures. Among their subjects were scenes from the gospels including the beheading of John the Baptist. This business expanded when around 1720 a metal stamping machine was invented, allowing clockmakers to replace the previous hand-carved wooden gears with metal ones. Eighteenth-century English table clocks were also embellished with moving figures, but these reflected the growing economic and cultural exchange between the empire and the Far East: Many of the clocks sported "oriental" characters.

With the introduction of the hairspring and balance wheel, moving figures were added to expensive one of a kind portable clocks and watches produced for the wealthy to give as gifts. In 1790, a Swiss watchmaker produced a musical automaton watch called the "Court Balancer." The clockface includes three tiny performers mounted in front of a painted landscape. While the two seated musicians move in time with the music, the central figure, a tightrope walker, repeatedly jumps and lands on the rope which gives slightly with the movement of the clock. This complex mechanism has separate clock and musical movements. Some of these watches displayed less wholesome scenes: rather than animals or musicians, they depicted lovers in various sexual poses, cavorting to the beat of the watch movement. The mechanical principles of clocks and music boxes were eventually adapted to freestanding automata. Those produced throughout the following centuries shared the public stage with puppet theaters, peep shows, waxwork figures (some of which were mechanized), and mechanical theaters. In addition to satisfying the public's unquenchable thirst for the curious and amusing, they demonstrated improvements in clock and watch making, as well as studies in acoustics and pneumatics.

AUTOMATA

By the mid-eighteenth century, automata, a term derivative of the Greek *automatos*—that which runs by itself—became associated with the Enlightenment concept of "Man the Machine." It may not seem that engineering is so closely related to philosophy, but in fact, the systematic observations of nature made in the sixteenth and seventeenth centuries by astronomers and physicists provided a catalyst for the Enlightenment concept of a knowable universe that could be described in mechanical terms. During the eighteenth century, the clockwork mechanism, now improved to an inspiring state of precision, became a popular analogy for the workings of nature among materialist philosophers like Julien Offray de La Mettrie and Baron d'Holbach who even characterized the human mind and body in this way. Those who produced automata could not have missed making a converse analogy: if people run like machines, then machines might be made to behave like people.

Androids

Like the clockwork mechanisms that inspired them, automata utilize certain mechanical elements in a consecutive manner such that the action of one component causes the movement of another. For instance, rods attached between the machine and its outer casing (body) cause it to move in a predetermined or programmed manner. The movement is cyclical, meaning that once the automaton is wound up, it performs the same pattern of motion repeatedly until it winds down.

Humanoid automata had come to be distinguished by the term, *android*: an automaton "in human form, which by well-disposed springs . . . performs many functions which resemble those of man" (Diderot and d'Alembert, *Encyclopédie*, 1751). For example, an early nineteenth-century advertisement for a display of automaton musicians in the Strand, London, notes that "The Androides will be found more curious than any Thing of the Kind ever before offered to the Public . . ." (Ord-Hume, 45). Many historians have called these androids the forerunners of robots; but we should not misunderstand the connection. It is not their resemblance to people or animals but their internal mechanisms that link them to modern robots. Androids were not just moving statues; they could actually perform some human behavior like writing, drawing, or playing a musical instrument. They operated according to a primitive form of programming or *artificial intelligence* (AI).

The androids owed their human-like movements to cam programming, a mechanical version of what is called in computing *Read Only Memory* or ROM. Cams are discs of varying sizes and shapes that rotate on a cylindrical shaft. Followers (rods) attached between the cam disks and parts like hands, make them move in a particular direction—up and down or back and forth. This combination cam and follower might be compared to computer programming and actuators of today's robots. Depending on the length of the follower and the shape, dimensions, and rotation speed of each cam, the object will automatically perform a specific, repeatable series of movements. These mechanisms created the impression of conscious behaviors like dancing, strumming stringed instruments, or writing, while hidden bellows caused air to be pushed through tubes to simulate singing, breathing, whistling, or even smoking. Both the memoirs of automaton makers and contemporary news reports indicate that very few androids were produced because of the time and expense of completing just one of them. Still, the exhibition of even a few of these made a number of mechanics famous.

Public exhibition was not only part of the entertainment culture, but an essential activity for scientists and inventors. Since the establishment of the first scientific societies in the mid-seventeenth century, those with new knowledge to share would present it to their peers for review, and also exhibit it to the general public. The exposure helped them to gain recognition and the respect, and especially to acquire funding to further their work. Promotion was a key part of this process for both inventor and the owner of the exhibition space. Advertisements for android displays were often quite lengthy explanations of what was new or different about the devices. Advertising was also used for fundraising. For instance, an ad from around 1820 invited the public to view demonstrations of a unique automaton. Subscriptions for the demonstrations, which could be purchased in advance from local booksellers, would fund the completion of the "paradoxical automaton." The author of the ad assures readers that it is not a hoax, and that unlike previous inventions, this one has a timer mechanism, so that rather than beginning its program as soon as it is wound up, the automaton can be preset to perform at a "period fixed, whether minutes, hours, weeks, months, or longer periods" (Ord-Hume, 47). The success of one of the most famous automaton makers of the eighteenth century can be attributed to the symbiotic relationship between inventor and exhibitor.

VAUCANSON

Early in his life, the Swiss inventor Jacques de Vaucanson (1709–1782) became interested in the sciences, mechanics, and particularly in simulating living beings. While studying with the Jesuits at Grenoble, he created angel automata. In 1735, after church authorities destroyed his workshop because they found his work morally objectionable, he left the order and moved to Paris. Although Vaucanson had been born into an aristocratic family, his inheritance was not significant enough to give him financial independence. Consequently, he devised a plan of borrowing enough money to complete one device, and use the proceeds from patronage and public exhibition to fund further research. By 1738, he had successfully produced two androids and an ingenious mechanical duck, which were all exhibited at the Hôtel de Longueville, Paris.

The first of these was a 6.5-foot-high figure of a faun (a mythical deity, top half man, bottom half goat) sitting on a rock and mounted on a 4.5-foot-high wooden pedestal. This "german-flute player" was already familiar to the public, for it was a copy of the marble statue by Coyzevaux on display in the Tuillerie Gardens. Spectators were impressed with the lifelike manner in which Vaucanson's android moved its head, lips, and fingers over the holes of the instrument. The main mechanism was not contained within the android's body, but in the pedestal, which housed a bellows to control the air; levers connected to the fingers, lips, and tongue by strings; and a barrel-shaped cam system connected to the pipes in the figure via a series of chains that provided the programming for a variety of tunes. A second android, mechanically similar to the German flute player, was a life-sized shepherd that played a pipe while keeping the beat with a snare drum.

Vaucanson's interest in simulating living organisms is evident in the invention most noted by his biographers—a mechanical duck. Mechanical animals were not unique in his time. For instance, a man named Maillard had already demonstrated a mechanical swimming swan to the Academy of Sciences in Paris (1731). However, Vaucanson's 400-piece duck had special abilities, which he describes in a letter to an Abbe De Fontaine. Here he notes that the duck, whose insides are visible to the spectator, ". . . stretches out its neck to take corn out of your hand; it swallows it, digests it, and discharges it digested by the usual passage. You see all the actions . . . to drive the food to its stomach, copied from Nature . . ." (Vaucanson, 21). He notes further that the duck drinks and makes a gurgling noise; and that no anatomist could "find anything wanting in the construction of its wings" (Vaucanson, 22), which move naturally.

During that year he presented a *memoir* (written explanation) and demonstration to the French Royal Academy of Sciences (f. 1666), where he proved beyond doubt that his devices were not a hoax, but legitimate simulations of the actions of living organisms. Thereafter the three automata were exhibited to appreciative audiences throughout Europe and Great Britain. Vaucanson had become so popular that Frederick II of Prussia offered him a position; but in 1741, he was appointed inspector of some silk factories to keep him in France.

There is no evidence that he ever made another automaton. Nevertheless, Vaucanson had accomplished his goal: the inventions had won him money, position, and respect. In 1746, he was admitted as a member of the Royal Academy of Sciences in Paris where he had first demonstrated his inventions; and later in his life, he became an inspector for that organization. After the trio of automata was exhibited in the Long Room of the opera house in the Haymarket, London, in 1742, his memoir was translated into

English and published by J. T. Desaguliers. It was also later translated into German and published in Hamburg (1747) and in Augsburg (1748).

Vaucanson was among the first mechanics to adapt his skill in automaton making to mechanization. During his tenure at the silk factories, he invented an apparatus to automate weaving. Thereafter, he made a number of other contributions to industry, including a metal cutting lathe and a boring mechanism. Late in life, he collected and displayed all of his inventions, which he bequeathed to France. A year after his death (d. 1782) the collection formed the basis of the Conservatoire National des Arts et Métiers. Unfortunately, his three famous automata were not in the collection, for he had sold them to exhibitors in 1741.

The duck had changed hands several times, disappearing and surfacing in various states of disrepair. In 1839 the remains were discovered in Berlin. A representative of a traveling museum purchased and restored the duck over a period of three years, and at some great expense. It was next displayed at la Scala in Milan, and in the 1840s, it was exhibited throughout Europe and reviewed by journalists and travel writers. The automaton maker Blaise Bontems (discussed below) is one of the last mechanics to handle Vaucanson's famous duck, which he acquired in 1863. After repairing it, Bontems sent it on tour under the protection of his nephew and apprentice, Seraphin. It was advertised in Valence as "Vaucanson's Famous Duck ... Repaired and Perfected by the Ingenious Mechanics, Monsieur Bontems of Paris." The display was a success, but what happened to the duck afterward is unknown. At the 1878 Universal Exhibition, Bontems was exhibiting only a copy of the duck; and in 1968 the descendents of Bontems still had the chain mechanism, but nothing more.

Although the whereabouts of the androids are unknown, Vaucanson's success in synthesizing musical sounds mechanically inspired others. For instance, in 1746, a man named Defrance exhibited his flute-playing shepherd and shepherdess that he advertised as being able to play 30 different aires. Musical automata were quite popular, although some were simply music boxes embellished with mechanical figures that only appeared to play instruments. Among the more realistic simulations of human behavior were those produced by the Swiss clock and watchmaker, Pierre Jaquet-Droz (1721–1790), one of the most highly respected mechanics of his time.

Like Vaucanson, Pierre first studied theology, but was more interested in mechanics. In fact, he was influenced by Vaucanson's duck, which was touring Europe when Pierre was about 17 years old. Eventually, he began making his own automata. By 1758 Pierre's reputation as a master mechanic got him an invitation to the Spanish Court, where he hoped for the patronage of the King. (Some accounts note that unfortunately he was not as well received by the Spanish Inquisition, and spent some time in jail there before returning to Switzerland.) With the assistance of his son, Henri-Louis, Jean-Frederic Laschot, and an apprentice, Henri Maillardet, the Jaquet-Droz shop produced a number of one-of-a-kind androids. The most famous of these are a musician, a draughtsman, and a scribe.

The idea of a writing automaton was not new. The earliest known Eurpoean writers were four prototypes produced by Friedrich von Knauss between 1753 and 1760. The fourth model, presented to Emperor Francis I of Germany, a tiny figure sitting atop a cast iron sphere, was about 1.9 meters high combined. A mechanism installed in the 80-centimeter wide sphere could be programmed to produce short sentences. The Jaquet-Droz writer, first exhibited in 1774, is more impressive because the mechanism is entirely encased within the child-sized figure. The "Scribe" represents a 3-year-old child sitting on a Louis XV stool. He rests his left hand on a writing desk, and holds a feather quill in his right hand. The whole arrangement is mounted on a small stand resembling a parquet floor. After great success at his home city, Chaux de Fords near Neuchâtel, the three automata were displayed in Geneva, Versailles, and London. After the mechanics settled into a successful watchmaking business in London, they sold off the trio to a French merchant based in Spain. The Jaquet-Droz androids fared better than Vaucanson's. After touring for decades, the three androids returned to Switzerland in 1906, and are still on display at the Musée des Arts et Histoire in Neuchâtel.

Henri Maillardet (1745 to c. 1800), one of three brothers in a family of Swiss clockmakers built a piano player that is undoubtedly inspired by the Clavecin-player produced by his mentor, Jaquet-Droz. His "musical lady" sits at a keyboard tapping her foot and moving her eyes, as her fingers depress the keys of the pianoforte. A concealed bellows pumps air into her upper body, making her appear to breathe while she performs one of sixteen programmed songs. Still a popular piece in touring exhibitions in the early nineteenth century, in 1820, the piano player was exhibited in London, promoted with advertisements lauding "her" many talents.

Maillardet is far better known now for his writer-draughtsman, no doubt also inspired by one produced by Jaquet-Droz. Maillardet's writer contains 96 brass cams just for the movements of the right hand, in addition to those that control the movement of the head, eyes, and left hand. However, like Vaucanson's automata, the cam mechanism is situated in a case below the figure. The size, shape, and relation of the cams constitute the instructions that allow the draughtsman to produce three poems and four drawings. The directions for movement are transferred from the cams to the followers for the limbs and hands via levers, rods, and pulleys. A motor in the pedestal slides the cams into place for each of the messages to be written.

Although the automaton eventually suffered from both fire and disuse, in 1928 its remains were donated to the Franklin Institute in Philadelphia, Pennsylvania, by the Brock family. They mistakenly presumed that it had been built by Maelzel, a French mechanist who had hoaxed an automaton chess player (actually operated by a small human inside the pedestal). However, the machinist at the Institute who restored the mechanism discovered it produced the works known to have been those of Maillardet's automaton. In fact, one of the poems is signed, "Ecrit par L'Automate de Maillardet" (written by the automaton of Maillardet). Today the restored writer-draughtsman can be viewed at the Institute's Science Museum.

Another popular category of automaton was the fortune teller or magician. These androids, dressed in the peaked cap and robes of a medieval sorcerer, were programmed to answer written questions. Most notable are the "Great Magician" and the "Little Magician" built by Maillardet. In both these devices, the platform is fitted with drawers and a number of question plates. The plates correspond to cam and clockwork mechanisms in the platform, which are visible in the Little Magician. When the "examiner" (spectator) pushes one of the question plates into the drawer, the magician responds by first gesturing as if he is considering the answer. He then gets up and points with a stick toward a little window above his head. The window then opens, revealing the answer plate that corresponds to the question plate. The magician sits down; and two tulips mounted on the platform bloom and close, indicating the end of the session. Two examples of the exchanges in the Great Magician are:

"What is the food of the soul?"—"Truth and Justice."

"What is the noblest purpose of science?"—"To reveal ignorance."

If the examiner closed the drawer without inserting a question, the magician would shake his head. If the examiner asked an impertinent (nonprogrammed) question, a skull would appear in the little window.

Fortune-telling machines continued to be a popular attraction into the twentieth century: In addition to weight scales and astrological "computers" that dispensed fortunes, automaton fortune tellers could occasionally be found in amusement arcades, department stores, and the lobbies of diners and movie theaters. When the spectator put a coin in the slot, the android would turn over some tarot cards, look from side to side as if contemplating them; and then drop a card with a fortune down a chute accessible by the spectator.

During the same period that Jaquet-Droz created programmable automata for a European audience, Japanese craftsmen turned out hundreds of *karakuri* (mechanical gadgets). The appearance of automata in the Far East has been attributed to the seventeenth-century Jesuit missionaries who brought them from Europe as gifts for the emperors in return for access to their lands. However, it is probable that some kind of figural automata or mechanical puppets were extant in China before the medieval period, for Chinese literary scholars have identified at least one fictional tale of a mechanical man from around C.E. 400. This may have been an extrapolation of the puppet theaters presented at royal courts. In any case there is a connection between the Asian puppet theaters and the automata that were made by Japanese mechanics.

KARAKURI

One of the earliest guides to mechanical devices in the East was the *Karakuri Zui (Illustrated Miscellany of Automata)* by Hanzo Yorinao (1796), which indicates that automata making was an established art before that time. Among the best known of all the Japanese automata is the tea carrying doll, one of many spring and weight-driven mechanical dolls, or *karakuri ningyo* produced during the Edo period in Japan (1600–1867). It is thought that the craft was initiated by the *Takeda-za* mechanical puppet theater in 1662 in the Dotonbori district of Osaka. The originals are unavailable, but they have been recreated a number of times in recent years from the Yorinao manual.

The tea carrying doll has been called the first autonomous mobile robot, since it possesses a self-contained power source, is self-propelled, and performs a useful movement automatically. As with most Japanese automata of the period, the doll's action requires the involvement of the viewer. In this case, the small automaton geisha figure holds a tray. When the "master" places a cup on the tray, the weight trips a mechanism, which sets the doll in motion. The guest then lifts the cup, drinks, and replaces the cup on the tray. This time, the figure makes a U-turn and walks back to the master. An adjustable spring makes it possible to send the doll either half or full length of a tatami (straw) mat.

POLITICAL ALLEGORIES

Figural automata sometimes represented political or social themes. "Tippu's Tiger," a double-figure wooden automaton depicting a crouching tiger preying on a British soldier, was allegedly produced around 1795 for Tippu

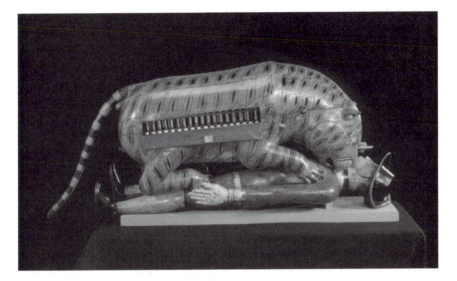

Tippu's Tiger, Mysore, India. Late eighteenth century. Victoria & Albert Museum, London/Art Resource, New York

Sultan, the ruler of Mysore, India. Because Tippu was known as the "Tiger of Mysore," the automaton has been interpreted as an allegory representing his feeling about the presence of the British East India Company in India.

The crank-driven organ that produces the sounds of the tiger and its screaming victim is supposed to have been added by a Frenchman in the sultan's employ. Some antiquarians have identified the soldier's uniform as Dutch, indicating that the soldier element may have been produced elsewhere. However, according to contemporary newspaper reports, a British officer named Munro who had been stationed in the area was dragged off by a tiger and later died in a hospital. This young man was the son of General Hector Munro, who had fought against Tippu. This suggests that the piece is an homage to the event. Ironically, after the fall of Seringapatam and the death of Tippu in 1799, the automaton was confiscated by the British and brought to London, where it was put on display in the museum of the East India Company. Today it resides in the Victoria and Albert Museum.

NINETEENTH-CENTURY AUTOMATA

In the nineteenth century, French toymakers in particular capitalized on industrial and economic expansion through France's long-established toy industry, which soon expanded to include music boxes, mechanical theaters, and automata under the heading of *bimbeloterie* (fancy goods). The center of nineteenth-century automaton making was the Marais, a district of Paris dotted with the shops of toymakers. Traditionally, toy making had been a small-scale enterprise, usually located in a one- or two-room shop in the artisan's home, with family members providing the labor. Automata "factories" were still located in the private homes of the craftsmen (or sometimes pre-Revolutionary villas), where different rooms were devoted to the production of heads, or arms, or movements, and where seamstresses (often the mechanics' wives) sewed clothing and hair.

Unlike the eighteenth-century automata that were painstakingly constructed individually over long periods of time, automata could now be made relatively quickly. Mechanics were able to take advantage of machined parts cut in multiples from sheets of metal. This reduced production costs—a savings passed along to customers. Automata were assembled in a production line fashion. It is interesting that despite the fact that they were still put together by human workers, the concept of the automated factory was already in place: We can think of the individual shop as a system in which every step of the process was carried out sequentially in one place.

Between 1820 and 1850 the number of French automata manufacturers had grown from 40 to 114; and after mid-century, with the help of exposure at international exhibitions (like the Crystal Palace of 1851, the Universal Exhibition in Paris of 1878, and the Columbian Exposition in Chicago in 1893) expanded even further. French automata were prized abroad and were exported in impressive numbers to the Middle East, India, and the Americas. Demand for the mass-produced automata came from a rising middle class. For a family of four whose yearly income was 1,500 francs in the 1870s, an automaton from the shop of Théroude at an average price of 1,000 francs was prohibitive. Nevertheless, they could purchase less costly pieces from other makers whose average price for one automaton was between 9 and 90 francs.

The era's atmosphere of spectacle was reflected in automata that represented theater or circus performers. Particularly popular were acrobats, musicians, and clowns. One of the most often-represented figures of the period was the clown "Little Tich." French colonialism also influenced the subject matter chosen for automata, reflected in exotic figures of the Orient, Middle East, and Africa. The demands of the market led in turn to further innovation and the expansion of the industry. The pace of production was so brisk that many makers never took the time to patent their designs, relying instead on the continued innovations to preserve their popularity with customers.

Automata businesses expanded and multiplied. Between 1855 and 1900, income from these items increased tenfold, from 7 million to 70 million francs. The intensity of competition is illustrated by the number of automaton styles copied, and patents for parts that were lifted and imitated. Just a few of the most popular firms were Théroude, Vichy, Roullet and Decamps, and Bontems.

Théroude

As a young man Alexandre Nicolas Théroude (1807 to c. 1885) worked with another toyseller; but in the 1830s he started his own business, importing and selling German toys and making his own. In 1840, affected by a slump in the sale of luxury items since the 1830 Revolution, he declared bankruptcy. When it was settled, he began to devote his business to the production and sale of mechanical toys and automata, and became one of the foremost figures in the industry. In 1849, Théroude was still making "ordinary" mechanical toys that retailed for between 3 and 35 francs; but he was already better known for his larger automata that sold for up to 1,200 francs each. Contemporary reporters noted both his mechanical skills—the realistic face of figures like the Child in the Cradle automaton that uttered "Mama" and "Papa" and the bleating sheep and goats that were covered with real animal skin.

Théroude was one of the most popular makers at the 1855 Exhibition where he was again applauded for the realism of his life-sized animal automata. Reporters covering the 1862 London Exhibition noted that Théroude adds to the "imitation of life" by placing his mechanisms inside his figures like his dancing couple, rather than hiding them in a pedestal. He evidently outdid himself at the Paris Exhibition of 1867, where he used a huge boulder as a stage for numerous automata including life-sized shepherd-musicians and monkeys, and a variety of smaller moving figures including a drum-beating rabbit. He was one of only a few automaton makers to enter the 1871 London Exhibition, held the same year as the Paris Commune; yet he once again stole the show. Unfortunately, economic difficulties in the aftermath of the Commune destroyed his business, and he declared bankruptcy again in 1878. Although a nephew whom he had trained carried on and opened his own automata-making business, Théroude himself never resumed his trade.

Vichy

Gustave Vichy (1839–1904) was the son of French clockmaker, Antoine Michel, and Geneviève Clemenceau, who also produced mechanical toys.

From the beginning of his career, Gustave, who had been trained as a clock-maker by his father, devoted himself almost entirely to producing automata. His wife, Marie Thérése Burger, a seamstress by trade, dressed his automata; and after the business expanded, she headed their large costuming shop. Although he had opened his shop in Paris in 1866, he did not participate in any of the exhibitions until the Universal Exhibition of 1878. There he was lauded for his lifelike figures. One visitor, noting the provocative looks and gestures of the Vichy automata, commented that if this exhibition were being held in the Middle Ages, Vichy "at the very least" would be excommunicated for his audacious behavior. From this time forward, the Vichy firm won numerous awards for their automata. In 1891 the French journal *La Nature* covered the now widely known Vichy firm as an example of contemporary manufacture.

Among the firm's popular subjects was the widely copied "Fin de Siecle" Moon head, in this case placed atop a human figure leaning on a column and smoking. The head slides side-to-side while the eyes roll. Another Vichy model featured two figures, a clown and a bespectacled man, poking their heads through an artist's paint palette. Although the bespectacled man held the pipe in his mouth, when it was lit, the smoke came out of the clown's mouth. Like many of the automata makers, Vichy also included in his repertoire the popular acrobats, dancers, exotic figures, and clowns. Near the end of the century, the operatic clown figure Pierrot was a popular feature. Vichy produced a Pierrot writer, and more than one version of him serenading the moon. By this time Gustave's son, Henry was a partner in the firm, and together they produced and exhibited numerous automata. A report from the Chicago Columbian Exhibition of 1893 notes that the successful Vichy firm employs 10 women, 15 men, and including its exports to the Orient, Europe, and America brings in 150,000 francs. In 1895 Henry incorporated Lioret phonograph mechanisms into several models, which were advertised as being able to speak, sing, and play musical instruments.

At this time, the United States was curtailing foreign imports, so Gustave and Henry made a failed attempt to set up a branch inside the United States. Gustave returned to France, where he kept up with the times by producing advertising automata with great success. He notes in one of his catalogues from the early 1900s that one of them won the Grand Prix at the Great Exhibition of 1900—the only award given for automata or mechanical toys. After Gustave's death, his widow sold the firm to his foreman, Auguste Triboulet for 20,000 francs. Although Triboulet did not introduce any new mechanical innovations, he successfully served the market by producing an eclectic assortment of amusing automata.

Roullet and Decamps

The firm of Roullet and Decamps is one of the longest-running automata businesses in France. Jean Roullet (1829–1907) worked in the upholstery business for the family of his wife, Laurence Francoise Midon, while he learned to be a mechanic (a term used to describe anyone who worked with machinery—from clocks and automata to industrial machinery). He opened his first workshop in 1866 primarily to produce and supply tools and stamped parts for other firms. His transition to automaton making was inspired by a request from a stone setter named Lamour that he produce an automaton gardener pushing a cart. Lamour had already invented a mechanism, but had made it by hand. They realized that using Roullet's mass-production techniques, they could put out dozens of these whimsical pieces for a fraction of the cost.

Roullet and Lamour became partners, producing "jouets automates" on the side while each continued in his own trade. They parted company in 1871, at which time Roullet set up shop in the hotel de Vigny at 10 Rue de Parc-Royal, Paris. Here his firm operated continuously for almost one hundred years. At the 1878 Exhibition, Roullet demonstrated the gardener along with a host of new models including a strutting peacock that spread its tail feathers, an egg-laying hen, a growling bear, and a team of horses pulling a carriage. He was awarded a bronze medal for his efforts. In 1879 Roullet's daughter Henriette married (Henri) Ernest Decamps, a mechanic and the foreman of Roullet's shop. The 1893 Chicago Exhibition report notes that the firm uses 16 different machine tools (although not electricity); and that it employs approximately 50 people, 30 of them men. It also notes that the prolific inventory of the Roullet firm is sold both locally and abroad, bringing in 200,000 francs per year. The report refers to the well-made automata sporting luxurious costumes, among them human figures and animals including a drum-playing bear. Roullet's automata of the period ranged in size from around 9 to 32 inches high, although the inventory included a 59-inch tall Black Flautist (1870). Like other automaton makers, Roullet's subjects were representative of the interests of Parisian society: exotic figures, monkeys performing human acts like smoking, acrobats, magicians, and especially clowns. He also produced versions of the Man-in-the-Moon theme, including a clown atop a moon face (1895). The piece produces three movements: A hat slides up and down the moon's head while from the clowns fingers, a spider drops down in front of his eyes on its web, and he licks his lips in anticipation. After the turn of the century Roullet retired, leaving the running of the firm to Henriette and Decamps, who took ownership after Roullet's death in 1907. By this time Decamps was producing electrically powered automata. One of them, a vignette with boxers, was described in the report of the 1908 Franco-British Exhibition as being particularly clever and well executed. An advertisement from 1910 boasts that the firm was a Grand Prix recipient for 1906–1908. At this time, he was also still producing impressive clockwork automata including his own version of "Little Tich," dancers, and acrobats.

Like Vichy, Decamps moved the focus of his business to advertising automata. After his death in 1908, the firm was assumed by his widow and children. His son Gaston produced his first electrically powered automaton display for a window in a Bon Marché store. This was a timely piece: a reenactment of Robert Peary's discovery of the North Pole, which he had achieved in April of that year. Around this time, the firm was also producing a line of mechanical toys representing various personalities, including Charlie Chaplin (1914).

After his brother Paul was killed in World War I (1915), Gaston took over the business and by the 1920s had bought out his mother's and sister's shares. Gaston, who had studied art and modeling, is remembered for the lifelike quality of his pieces as well as for their wit. His Santa Claus Flying over the Place de Vosges, Paris, accompanied by several animals was a hit at the 1925 Paris Exhibition of Decorative Arts. It included a polar bear that pours a cup of cocoa and drinks it down. Gaston also acknowledged the Jazz Era with a life-sized black jazz band that he produced for a store window. He died in 1972. By that time the family had been evicted from the Rue de Parc as a result of real estate and redevelopment conflicts. However, they continued in other locations and at the end of the twentieth century, the firm was being run by Gaston's daughter, Cosette and her husband Georges Bellancourt.

Bontems

Blaise Bontems (1814–1881) whose early training was stuffing birds became one of the foremost producers of musical bird automata. He was allegedly inspired to his craft while he was an apprentice clockmaker; when a customer brought in a musical snuff-box to be repaired. According to his biographers, Bontems was disappointed that the whistle was unnatural, and after spending some time in a forest learning the nightingale song, he modified the music box mechanism in such a way that it made a more authentic sound. From this experience, he was inspired to make others, which he sold at rather high prices. By the mid-1800s, Bontems had his own business and became famous locally for his singing bird automata, ornate dioramas of birds in trees and arbors that produced realistic bird songs. Jurors at the 1851 London Exhibition referred to his "ingenious drawing room ornaments containing automaton birds ... more for adults than children ..." (Bailly, 44).

A decade later, Bontems' work was admired around the world; 90 percent of his product was made for export. His subjects included monkeys, organ players, and "negro" smokers, but he was particularly known for his bird automata, which were lauded for the realistic quality of their songs, including those of canaries, finches, nightingales, and blackbirds. A reporter at the 1867 Exhibition described "the soul" of the singing birds as a clockwork mechanism, a pinned cylinder like a music box, a bellows system, and whistle with a tiny piston to modulate the birdcall. An 1878 exhibition report notes that his pieces not only displayed believable birdcalls, but that the wings, tails, and bodies also moved in a realistic way. Bontems had always wanted to produce a human-speaking automaton; and though he once told a contemporary that he had produced one that could speak his apprentice's name and a few other syllabics, there is no evidence that he ever finished it.

Bontems' records reveal the collaboration that took place among these otherwise competitive mechanics to further their own businesses. For instance, his records show that Théroude mounted some pieces for him; and that another mechanic named Saugrin produced some of his musicians (although it is not clear whether Saugrin produced the mechanical or the figural parts).

After his death his son, Charles Jules took over the successful firm, and exhibited successfully at the 1893 Chicago Exhibition. A 1900 Paris Exhibition report noted the Bontems bird automata range in complexity, and in price from 90 to 750 francs. During this period, the firm employed 12 men and eight women, and was producing 400 pieces a year. Jules was succeeded by his sons, and the business stayed in the family until 1966 when Lucien Bontems died. At this point, it was sold to the Swiss firm, Reuge, which even today makes a line of Bontems-inspired bird boxes.

Advertising

Entertainment automata did not die out, but over the next century were transformed into a multibillion-dollar toy and entertainment industry, discussed in Part III. During the last decades of the nineteenth century, automata were placed in the windows of the newly developed department stores as advertising gimmicks. The first of these were similar to the small drawing room automata, with the addition of the name of the company and product they advertised displayed on their bases. Gradually, they were built larger to fill display windows, to compete with other shop owners, and to attract the curious from further down the sidewalk. Like those made by Vichy and Decamps, they could be extremely complex scenes. By the early twentieth century, clockwork mechanisms, which needed to be continually wound during the course of the day, were replaced by electric motors, which ran as long as they were plugged in. Their descendants are the mechanized dioramas displayed in the windows of department chains like Macy's at Christmastime.

INDUSTRIALIZATION

Improvements in regulators for steam engines during the early part of the nineteenth century made it possible for people to think of automatic machines not just as amusing devices, but as the basis of profitable businesses and the foundation of national economies. Political and economic expansion provided the motives for advances in communication, transportation, and especially the mass production of consumer goods and military instruments and weaponry. It also created a fierce competition among nations that recognized the need to build strong infrastructures, well-equipped military and competitive import and export capabilities. As with the production of ushebti thousands of years earlier, machine parts and the resulting manufactured goods were standardized to promote the efficiency and speed of the mass-production process.

Mechanization to Automation

The textile industry was one of the first to incorporate control and programming theories, and the first engineers of the industry were clock and automata makers. Like Vaucanson before him, the technologist Joseph Marie Jacquard (1752–1834) contributed to the first advanced weaving system. Under the patronage of Napoleon, Jacquard introduced a punch-card system for controlling the cams that in turn controlled the pattern output. Chains of these pattern cards were combined to produce intricate patterns. These were the first modern instance of stored programming for industrial purposes: Each hole or absence of a hole can be compared with the "on/off" switching principle of modern computing (see Chapter 4). The Jacquard loom was in wide use in the textile industry by the second decade of the nineteenth century.

It soon became apparent that the advantages of automation to textile manufacturing could be extended to other industries. Mathematician Charles Babbage planned to use Jacquard's system in the Analytic and Difference engines he was developing to compute navigational tables. Though Babbage had to abandon his projects, others were eventually able to adapt the punch-card system for actuarial and military activities. For instance, the 1890 census was tabulated using a punch-card computer designed by Herman Hollerith.

Mechanical improvements in the printing trades are illustrative of the way that automation altered the character of an industry. Taking advantage of improvements in the steam engine and controllers, the communications industry expanded the accessibility of information while simultaneously creating demand for it. It also transformed what were traditionally separate cottage industries into an integrated factory system. The mechanization of paper production made it possible to print, distribute, and purchase paper at a lower cost, a savings passed on to consumers in cheap print. A steam-driven papermaking machine invented by Nicholas Robert and improved upon by Bryan Donkin in the early part of the nineteenth century automated several processes formerly done by hand: combining pulp and water, coloring, sizing, and distributing the fibers uniformly across the paper form. The pressing and drying operation was also mechanized and powered by steam engines. By 1851, almost 200 of these machines had been installed in mills. Innovations in printing presses were made concurrently with advances in papermaking.

In 1812 Frederick Koenig produced the first twin-cylinder steam-powered press. Two years later his invention was installed at the *London Times,* where output was increased by 400 percent to 1100 printed sheets per hour. Consequently, distribution of the *Times* expanded to over 30,000 papers per day by the end of the 1830s. The subsequent invention of the rotary cylinder press by Richard Hoe led to substantial increases in the production and distribution of newspapers. His machine, installed in the Philadelphia *Public Ledger* in 1847, produced 8,000 newspapers per hour. This innovation was followed by the invention of the Web press by William Bullock in 1865. Bullock's press allowed for printing from a continuous roll of paper. Hoe and his partner Stephen Tucker added the capability to print on both sides of the paper roll at once. Their "perfecting" Web press, installed at the *New York Tribune* in 1871, could print 18,000 entire newspapers per hour. In 1875 Tucker patented a rotating folding cylinder that automatically folded the papers as they came off the press.

Despite these innovations in papermaking and printing, it is the mechanization of type-casting that is often hailed as the most important contribution to the mass publishing industry since Gutenberg developed moveable type. During the 1870s the German-born inventor, Ottmar Mergenthaler (1854–1899) invented a machine that solved the dual problems of typesetting and casting. He was granted a patent in 1884, and by 1886 one of his machines had been installed at the *New York Tribune*. It was a vast improvement on the slow and cumbersome process of placing individual characters into long casting sticks and transferring them by hand to the printing page form. The machine consisted of a keyboard and a casting unit containing brass that had been heated to 500 degrees Fahrenheit. The unit was first adjusted for the size of type and length of line. As the operator typed characters on the keyboard, *slugs* or *matrices* (characters) were stamped out. These moved from the casting unit along a conveyor belt to a *composing stick* (assembler box) to form an entire line of type—hence the name, *Linotype*. The advantage was that one operator could perform the work of several people, and in less time.

Just a year after the Mergenthaler machine entered the publishing industry, the American-born Tolbert Lanston invented a *Monotype* machine. Lanston's system was based on precise mathematical calculations for the size and spacing of individual characters. As with Mergenthaler's system, the operator typed on the keyboard. However, in this case tiny perforations in the shape of the characters were made on a paper cylinder. This was used to cast the more accurately sized and spaced slugs.

The transformation from cottage industry to factory system was not only a consequence of technological innovation but of social and economic philosophy. By the early nineteenth century, social scientists and economic theorists trying to identify methods for increasing national wealth through industry had argued that self-regulating machines could simultaneously increase output and solve the problem of inefficiency in human workers. To illustrate the point they drew on the now familiar man–machine analogy articulated during the Enlightenment. Andrew Ure saw the ideal factory as

> A vast automaton composed of various mechanical and intellectual organs, acting in uninterrupted concert for the production of a common object, all of them being subordinated to a self-regulating, moving force. (Ure, 1835, 13)

Ure saw an economic advantage in replacing skilled labor with a more docile (and cheaper) workforce of women and children. They would tend the more reliable machinery that in effect managed the workers. His contemporary, the mathematician Charles Babbage (who was working on his mechanical calculators) wrote, "One great advantage which we may derive from machinery, is from the check which it affords against the inattention, the idleness, or the dishonesty of human agents" (Babbage in Morus, 157). A new class, industrialists, welcomed the idea of sophisticated, self-regulating machinery that would allow businesses to compete in a growing international marketplace through local distribution and export of cheap, quickly made, mass-produced goods. They pictured every member of society from machinists, owners, and managers, to workers and consumers contributing to the smooth operation of society, an operation often characterized in mechanical terms.

Scientific Management

Echoing the sentiments of Babbage and his contemporaries earlier in the century, Frederick Winslow Taylor (1859–1915) promoted to American business and industry the idea of the workplace as a well-oiled machine. Inspired by his observation of disorganization and nonmotivated workers at the Midville Steel Company plant where he worked as machinist, then as foreman, Taylor devised a new system based on his own time and motion studies. His goal was "a self-perpetuating system of management based on the efficient utilization of scientific knowledge." Tools were rearranged for worker convenience, and workers were paid piece-rate rather than hourly wages, set according to his calculations of how much work could be done in an average day. The new incentives were meant to be mutually profitable to worker and employer.

Beginning in the 1880s, Taylor popularized his "scientific management" theories for workshop efficiency through his books and public speeches. These principles supported the national initiative in economic competition through industrial expansion, which he promoted in his address to a congressional subcommittee hearing on the topic in 1911. He promoted methods already known by the term *driving*: piece-rate incentives, speed-ups, and firing those who could not meet quota. Taylor held a romantic vision of the efficient workplace as one that would assist workers in returning to their historical roles as artisans, turning out high-quality goods and distinguishing themselves from less productive employees, thus increasing both their wages and their status. Thus he and other proponents of modern industry coopted the romanticism of Thoreau (who thought conversely that industrial machinery destroyed individual creativity and sensitivity). In 1913, Taylor's *The Principles of Scientific Management* made the list of the most popular books on sales and efficiency among businessmen, clerks, managers, and foremen.

Taylor's protégés introduced the principles of scientific management into factories, department stores, railroad companies, banks, publishing houses, and construction companies. *Taylorism* is at the heart of Frank and Lillian Gilbreth's workplace efficiency studies. The Gilbreths expanded on Taylor's time and motion studies by pioneering the use of slow motion cinematography to document the amount of time, space, and effort taken to do specific movements in the course of completing a task. It was Lillian Gilbreth who devised the system still popular in kitchen design, in which food storage, preparation, and cleanup are laid out in a triangular pattern to eliminate wasted movement.

Although they were promoted as a method of making workers' lives easier by streamlining their physical movements, the main goal of efficiency programs was to make the employee's output more profitable to the company. The owners knew it, the workers knew it, as did social critics, whose critique of the modern factory system made its way into popular culture. For instance, Charlie Chaplin parodied efficiency programs in the American Depression-era film, *Modern Times* (1936), which includes a scene with a machine feeding employees so that they can keep working through lunch hour. If Taylorism as a labor philosophy was repugnant to social critics, many progressives embraced the idea that engineers themselves were good for society. This was a result of the rise of industrialized nations.

Engineers

Between the 1820s and the 1920s, the mechanical engineering profession was transformed from one run by self-taught mechanics or those apprenticed on the job to a cadre of college-educated experts. By the middle of the nineteenth century, the training of mechanics and engineers was an important initiative in many countries. Social theorists in England began arguing for upgrading its own technical education to remain competitive; and by the 1850s, Britain had begun to establish its "Red Brick" school system, essentially to train engineers and managers for coalmining, textile, and its new chemical and steelmaking industries to compete internationally. Once Germany had consolidated itself in the 1870s it immediately established polytechnics or universities of technology. The Japanese, after centuries of isolation, embraced the industrialization and modernization of their country, and by the end of the century had the largest electrical engineering institution in the world. The United States had been establishing land grant schools that focused on training in agriculture and the mechanical arts, and in 1890, the government provided for an annual stipend for those schools.

The importance of scientific and technical training in the United States is illustrated in the establishment of the Massachusetts Institute of Technology in 1865 and the research center, Johns Hopkins University in Baltimore, Maryland (f. 1876). By the turn of the century, the clash between what had been referred to as "shop culture" (learning within the environment) and the university educated technical specialist ended in many gains for the professionals. Entrance to technical societies even became restricted to college-trained professionals.

Technocratic Movement

During the first decades of the twentieth century, progressive intellectuals identified the engineer, who had gained respect as a scientifically trained expert, as the best manager of society's future. In the United States, intellectuals revived the Enlightenment cause of rationality as a basis for their position. Ironically, it might seem, Taylorism was embraced by technocrats who were willing to forego a little democracy for a lot of efficiency and productivity. Taylor's own chief assistant, Henry L. Gant, associated his mentor's philosophy with the Protestant work ethic. According to those like Gant, technologists could end poverty, class conflicts, and prejudice. The "myth of the engineer" emerged to justify their position at the helm of progress. More importantly though, they saw ideologically neutral engineers as owing nothing to political bosses, and therefore the most likely candidates to ensure that the progressive goals of efficiency and social uplift were accomplished by overcoming the fundamental economic causes of those conflicts through technological progress. With the skills, the rational judgment to understand the industrial complex on a deeper level than those whose interest was purely their own political or economic gain, experts were seen the likely choice for planners and managers. While Enlightenment philosophers had used rational thought to discover and classify the elements and operation of the universe, modern rationalists put their systematic thinking to the project of an efficient, productive, and fair use of that universe. The long-sought-after efficiency and productivity could be had through standardization. Machines that could repeat the same precise movement, provided there was a continued power source, were practicable if the large numbers of products were of standard sizes, colors, weights.

Other factors supported the move toward automation: the cost of steel fell dramatically in the late nineteenth century so that the focus shifted to perfecting the products refined from raw materials. The level of electrically powered machinery had grown from 4 percent in 1899 to 30 percent in

Steam Men

In August 1868, the popular story magazine *Beadle's Dime Novels* ran, *The Steam Man of the Prairies*. It was the one and only invention story written by Edward S. Ellis, then a superintendent of schools in New Jersey, and a prolific dime novel writer who had also authored 50 books on American History. The tale revolves around Johnny Brainerd, a "humpbacked" young man from St. Louis, Missouri, who invents a steam-driven mechanical man that pulls a cart. Johnny takes the contraption to seek adventure in the "Wild West," where typical dime novel excitement ensues. Steam man stories reflected American's enthusiasm for technological innovation, and industrial and geographic expansion. However, they were more directly inspired by a series of newspaper reports that began to appear before mid-century about similar inventions. One of the earliest of these was a blurb in *Scientific American* in 1849. The reporter announced that a mechanic in Russia had assembled a giant metal statue and attached it to the front of a locomotive engine. The blurb was reprinted as filler in the *New York Daily Times* on March 5, 1853. After the Civil War, a team of mechanics in Newark, New Jersey, filed a patent for a mechanical steam-powered walking man that pulled a cart. Zadoc P. Dederick and Isaac Grass demonstrated their invention in January, and on March 24, 1868, they were granted patent #75,874 for their "Steam Carriage." The men had spent $2000 on the prototype, but told reporters that their objective was to mass-produce the contraption at a unit price of $300. The carriage was meant to be used as a kind of taxi.

In August 1868, *Beadle's Dime Novels* ran Ellis' story in issue #45. Beadle reprinted the Ellis' story before publisher Frank Tousey bootlegged it and had it rewritten by Harry Enton as *Steam Man of the Plains*. For the next 50 years, reports of inventions of steam men (and subsequently electric men) appeared in newspapers from Ontario, Canada, to the Ohio Valley in the United States. They shared the spotlight with spin-offs and reprints of Ellis' dime novel, and even hoaxed reports of steam men. By the early decades of the twentieth century, the contraptions had been exhibited to thousands of people across the country. Although the inventions ended up in junkshops, and the dime novel readers' interested faded, the steam and electric men phenomenon had contributed to the emergence of the mechanical man as an icon of the future.

1914 when World War I began, to 75 percent by 1929. Together the availability of materials, the institution of electricity as a power source, and the economic incentive to manufacture products for export, domestic, and military use provided an incentive to develop automatic machines. This motivated improvements in control engineering.

Control Engineering

Control originally referred to regulating the speed and general movement of simple machines, for instance the flywheel attached to the potter's wheel and the bow drill, and later to mechanizing the on–off procedure in machinery via switches and toggles. Eventually, control engineering was understood as the regulation of speed, direction, or force through feedback and *servomechanisms*. The Flyball governor, a speed regulator invented by James Watt (1788), was one of the first servomechanisms. Instead of having a human operator manipulate the throttle of the steam engine to increase or reduce steam pressure, the flyball was set in a spinning motion. The faster it spun, the less steam was admitted into the cylinder. Almost a century later, Sir James Maxwell published his thesis "On Governors" (1868) in which he formalized the principles of control engineering. In 1893, Christopher Spencer filed a patent for a cam-programmed screw-fabricator. Factory machines could handle more work and for more consecutive hours than people, but still required the involvement of humans to operate them. By the end of the nineteenth century, improvements in servomechanisms for navigation and manufacturing provided the first opportunity to move toward the true automation of machines, and the removal of the human worker from the process. The robot emerged from this amalgam of technical innovation and socio-intellectual transformation.

Part II

THE EARLY YEARS

3

Into the Factory: The First Robots

I shall show an automaton which left to itself, will act as though possessed of reason and without any willful control from the outside. Whatever be the practical possibilities of such an achievement, it will mark the beginning of a new epoch in mechanics.

—Nikola Tesla (1898)

In 1898 the inventor, Nikola Tesla demonstrated a model for a remote-controlled submarine at Madison Square Garden in New York City. Allegedly, when a reporter suggested that the machine could carry and detonate torpedoes undetected, Tesla told him, "You do not see there a wireless torpedo, you see there the first of a race of *robots*, mechanical men which will do the laborious work of the human race." Many of Tesla's predictions were met with laughter during his lifetime; but by the 1920s, engineers were working in earnest on automation to serve the expanding needs of the military and the demands of consumer manufacturing. Their goal was to build industrial machines that could operate efficiently for long periods by self-correcting any irregularities in functions like velocity, force, position, or acceleration. This was accomplished by producing better sensing mechanisms to send error signals or *feedback* between the input and output parts of the device; and *servomechanisms* to correct any error (difference) between the intended and actual output. An indicator of the rising importance of control engineering is the proportional increase in the sale of controllers over other scientific instruments—from 8 percent in 1923 to 32 percent in 1935. By that time, approximately 75,000 controllers had been sold to industry (Bennett, 1993, 28). Advances in control engineering during the years between the two world wars made it possible to build machines that fulfilled Tesla's vision.

SHOWING OFF

In keeping with the tradition of holding public demonstrations of innovations, "mechanical men" appeared at trade shows and world fairs, amusing the crowds as they showed off the latest in electromechanical engineering. In 1928, influenced by publicity from Karel Čapek's play, the aluminum robot that opened the Model Engineering Exhibition in Gomshall, England, had "R.U.R" engraved in its chest. The Gomshall robot was able to sit and stand via electric motors; and an electrical gear enabled it to hear and answer questions.

In the same year, a huge golden-colored pneumatically controlled robot was exhibited at an exposition held in Kyoto, Japan. This entertainment robot was called GAKUTENSOKU—Japanese for "Learning from the Rules of Nature." During the next decade, exhibition robots opened a number of public events, including the Texas Centennial Exposition in Dallas in 1936.

Westinghouse Electric Company often used robot mascots to show off their innovations. In 1929, J. L. McCoy, a Westinghouse representative, demonstrated a predecessor of the remote presence home robots now on the market. The cardboard TELEVOX, designed by Joseph Barnett, responded to orders via a telephone connection between a circuit board on the stage and the one in its torso.

Although TELEVOX was billed as a "household robot," the "chores" it managed were minimal: it could turn on appliance switches, but not operate the machines. Its main purpose was to demonstrate a new type of controller that was being installed to monitor water levels in reservoirs. The company next exhibited WILLIE VOCALITE, which spoke in a deep baritone voice and could sit up and fire a gun. WILLIE attracted the crowds to view other new products. In 1939, Westinghouse exhibited a new robot called ELEKTRO at the New York World's Fair. The 7-foot-tall robot, powered by electric cables attached to its heels, weaved through the hall with a robot dog SPARKO, talking to visitors and answering questions via speech synthesis.

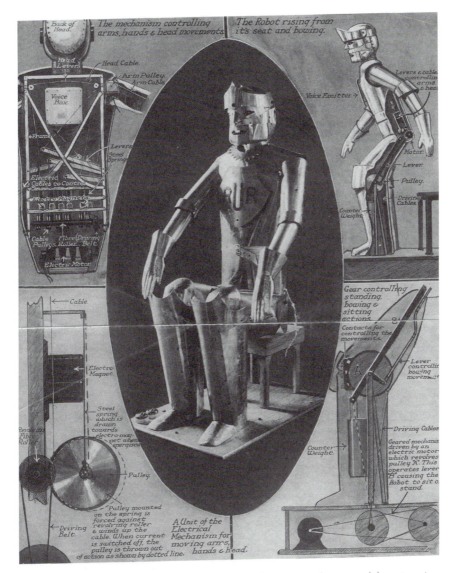

Mechanical Man dubbed the "First British Robot" demonstrated at a model engineering exhibition, Royal Horticultural Hall, Gomshall, England, 1928. Courtesy of Photofest

Exhibition robots could generate enthusiasm for the latest in electrome-chanical control engineering, but they would be of little use on an assembly line. On the other hand, factory machines like those that filled and capped bottles could process enormous batches of product quickly but were not very versatile. It was costly and time consuming to retool them for different

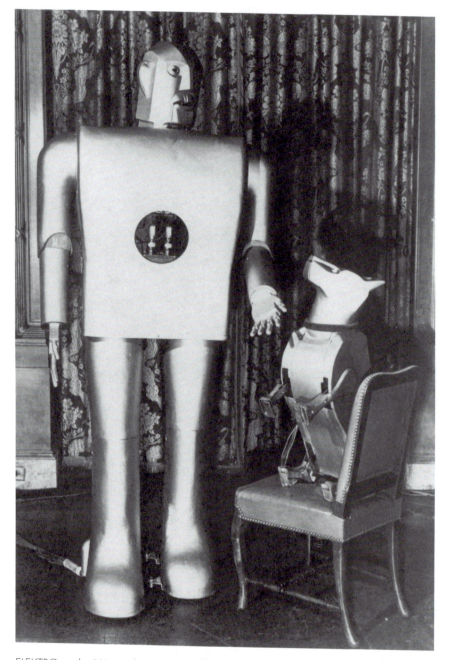

ELEKTRO at the Westinghouse exhibit, New York World's Fair, 1939, shown with his mechanical dog Sparko. Courtesy of Photofest

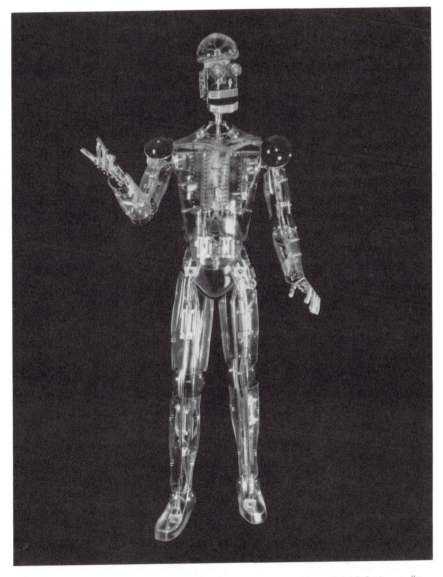

Robots continued to be corporate trade show attractions. Here GORDON, a talking robot explains computer software at AT&T InfoQuest Center, New York, 1986. Courtesy of Photofest

tasks or to perform sequential operations. Nor were they able to self-correct if something went wrong. During the next 20 years, researchers working to adapt control engineering to the factory produced the first true robots. The earliest designs for industrial robots were introduced before the United

States entered the Second World War. These jointed *manipulators*, some of them modeled on human shoulder-arm-wrist kinetics, could replicate human motions like pulling, pushing, pressing, and lifting via cam and switch programming. The main performance goals were repeatability, precision, adaptability, and safety.

In 1938, Willard V. Pollard filed a patent application for the first of these arms, "Position Controlling Apparatus." It had a primitive electronic controller comprising *Thyrotron* tubes that recorded position on a magnetic drum. Pollard's design was a two-arm affair with a flexible shoulder and wrist. Pneumatic cylinders and motors powered its six axes of motion, providing it impressive dexterity for such an early design. A year later, Harold A. Roselund designed an automatic spray-painting arm with wrist movement. This more compact manipulator had counterbalances at the shoulder and elbow, and was driven by a large multitrack drum cam system. The DeVilbiss Company, one of the first U.S. robot suppliers, received the patent for Roselund's design, "Means for Moving Spray Guns or Other Devices through Predetermined Paths." Unfortunately, the large, unwieldy memory drums made reprogramming the arms time-consuming and difficult. Further improvements in both sensors and controllers were necessary before truly useful robotic manipulators could enter the factory. During the World War II era, research to increase the precision of automatic weapons and aircraft, and to improve computing for actuary, communications, and code breaking provided two theoretical bases for intelligent machine control: *cybernetics* and binary computing.

CYBERNETICS

A group of researchers working on automatic weapons systems adopted the human central nervous system as a control paradigm. In 1943 Arturo Rosenblueth, Norbert Wiener, and Julian Bigelow coined the term "cybernetics" (from the Greek word for "steersman"), to equate data processing with the natural feedback mechanism that stabilizes the temperature of warm-blooded animals. An animal continually communicates its sensorial experience to its central nervous system as automatic, involuntary feedback to regulate processes like temperature, respiration, circulation, and digestion, and responses like fright or fatigue. Other types of feedback are to some extent under conscious control. For instance, when a person intentionally reaches for an object, information about the position of his/her arm and hand is fed continuously back to the brain both by the eyes and by position-sensing nerves in the arm. The brain uses this position information to guide the hand to the object and complete the desired movement.

These researchers theorized that the organism self-regulates by scanning for the level of success (or failure) of a certain behavior, and uses the result as the basis for modifying future behavior. They argued that mechanical devices could operate in the same way. A number of machines are regulated with sensors. When an engine's speed exceeds its preset limit, a governor reduces the supply of fuel, thus decreasing the speed; and electronic control systems like voltage regulators employ feedback extensively. The most-often described example of sensor regulation is the household thermostat. In this closed-loop control system, a sensor—in this case a metal wire—bends as the room temperature changes. The device self-adjusts the heater up or down according to the desired temperature setting.

As simple as a thermostat is, they reasoned, an industrial machine could self-regulate using the same internal logic: A message about the difference between the commanded movement and a machine's actual position or action could be sent back to its processor, and a servomotor would correct the difference. Cybernetic theory was adapted as a solution to the chronic instability of feedback mechanisms; and incorporated into stabilizers for radar-controlled antiaircraft guns. After the war, in 1946, neurophysiologist Warren McCulloch gathered together mathematicians, computer engineers, physiologists, psychologists, sociologists, and anthropologists at the first of the Macy conferences on cybernetics in New York. Together they would forge an interdisciplinary expedition into the realm of machine operation using biological systems as its paradigm.

In his book, *Cybernetics: Or The Science Of Control and Communication in the Animal and the Machine* (1948), Norbert Wiener, who had participated in earlier gatherings on the topic at Princeton (1943 and 1944), reiterated the idea that inanimate systems could simulate biological and social systems using improved sensors. Such improvements made it possible for engineers to adapt cybernetics theory to industrial machines. Even so, servo controllers were still inadequate to the long-term goals of automation. However, wartime efforts to improve encryption systems and weaponry resulted in the expansion of computer science. In turn, computers would prove invaluable to the automation of industrial machinery.

COMPUTERS

As we have seen with the Antikythera mechanism, the concept of mechanical calculation is an ancient one. Mathematician Wilhelm Schickard made one of the first modern attempts to produce an arithmetical machine that adds, subtracts, multiplies, and divides in 1623. Unfortunately, only Schickard's original schematic remains. In the 1640s, Blaise Pascal produced a shoebox-size dial-type calculator that could handle numbers up to 999,999,999; however, the design was too expensive to reproduce for practical use. Gottfried Wilhelm von Leibniz also failed in his attempt to build a calculating wheel in 1673. The "Leibniz Wheel" worked, but was too simple to be of much use to scientists. Leibniz' real contribution to computing was base-2 arithmetic, or *binary logic*, based on the concept that everything in the universe is either existent or nonexistent. As far as we know, except for the ancient Hebrew cabalists who described creation in terms of opposites (back and forth, alive and dead), Leibniz was the first to streamline calculation by replacing our base-10 counting system with one that uses only zeros and ones.

Leibniz left only preliminary notes behind; but in the mid-nineteenth century, mathematician George Boole expanded upon Leibniz' idea. His system, later called "Boolean Algebra" in his honor, was published in 1854 as *The Laws of Thought*. Boole referred to 1s as "universes" and 0 as "nothingness." Things either are or are not. Meanwhile, the British government had funded mathematician Charles Babbage to produce what he called an Analytic Engine—a football-field-sized steam-driven calculator with a typesetting feature to print out answers. Babbage had been active in promoting funding for technology development in Britain, and contributed both social commentary and practical schematics. In 1822, he had built a small-scale model of a difference engine. It was for this model that he received funding for the Analytic Engine, which he worked on with Lord Byron's daughter, Lady Lovelace.

Babbage intended that his machine would streamline and render the tables used in ballistics and navigation error proof. He was inspired by working models of the calculating machines designed by Leibniz and Pascal, and the punch-card sequences designed by Jacquard for his textile loom. Unfortunately, after 20 years of work, Babbage had still not completed the machine and he lost his funding. He continued for a while using his own funds, though he never completed his project. Nevertheless, Babbage's project inspired a basic tenet of modern computing: conditional or branching logic. This is the "IF/THEN" source of modern rules-based computer programming: If one condition exists (or does not), then another condition must (or must not) exist.

By the end of the nineteenth century, Herman Hollerith built a practical machine that used the principles of these pioneers. He incorporated Jacquard's punch-card system with the binary system to produce a desk-sized tabulator. The machine was first used to count the 1890 census, which it completed in only two years. The relative speed with which the tabulator read demographic information punched into the cards is evident in the fact that the previous census was still being calculated by hand when Hollerith finished his machine in 1880. The value of his invention was immediately recognized and his new company, International Business Machines (IBM) remained a leader in tabulation and computing devices throughout the twentieth century.

By the 1920s researchers were trying to produce calculators that could do more complicated mathematics, for instance, solving differential equations. For instance, at MIT, Vannevar Bush produced a differential analyzer that improved mechanically on the one designed by Lord Kelvin in the late nineteenth century. Kelvin's idea was to solve the equations in stages, but the technology of his day was not advanced enough to build an adequate gear system. In 1930, Bush was able to build a practical model of the integrator. From that time, both technological innovation and the perceived need for high-speed automatic calculations pushed computing forward significantly.

During the late 1930s, Claude Shannon, then an assistant at MIT, invented a calculator composed of an electric switching system that demonstrated a practical application of binary logic (the analog switches, which could be only on or off are analogous to the 0s and 1s of digital switching). The first truly electronic computers were produced during World War II. They were hundreds and in some cases thousands of times faster than the relay switch computers developed by Claude Shannon. In 1940, John Atanasoff and Clifford Berry designed ABC, the first all-electronic, nonprogrammable computer, and the British computer research group Ultra produced ROBINSON, an electromechanical relay computer used to crack German code. In Germany, in 1941, Konrad Zuse, working on computerizing aircraft controllers, developed the Z-3, the first operational programmable digital computer, programmed by Arnold Fast. The COLOSSUS series, initiated in 1944 by Britain's Ultra computer team to decipher highly complex German codes, could calculate up to one thousand times faster than their own relay switch computers.

Howard Aiken produced a machine that calculated twelve times faster than a human being did. His graduate school at Harvard was uninterested in the design that had arisen from his PhD dissertation work. IBM, committed to encouraging research beyond the level of the tabulating machines it was producing, supported Aiken's project; and Aiken completed his MARK I in 1945. It was a mechanical computer, approximately 8 feet tall and 50 feet long with three quarters of a million parts and hundreds of miles of wiring. It read its instructions from punched paper tape, and its calculating power came from vacuum tubes; but it was the first programmable computer built by an American.

The model for the modern, general-purpose computer was ENIAC, conceived by J. P. Eckert and John Mauchly at the University of Pennsylvania Moore School of Electrical Engineering as an improvement on Vannevar Bush's differential analyzer. Constructed for the U.S. Department of the Army to calculate artillery tables, ENIAC could produce 20,000 multiplications per minute, and because it was capable of subroutines, was far more versatile than computers like the MARK I. Unfortunately, the army continued to increase its demands for what the machine should do, and although ENIAC was a success, the project that had begun as a contribution to the war effort was not completed until 1945.

John von Neumann's essay on the stored-program computer established the more efficient processor/memory protocol that became the standard for computing. Von Neumann realized that both data and the instructions for manipulating it could be stored in the same circuitry—that he dubbed the machine's "memory." Calculations would take place in the central processing unit (CPU). The first machine built on this principle was the Electronic Discrete Variable Computer (EDVAC), created at the Institute for Advanced Study at the University of Pennsylvania in 1946–1951 for engineering applications.

The introduction of transistors into computers beginning in the mid-1950s reduced their physical size and increased performance significantly. This meant that computing could be applied to a wider range of applications, including automation. During the next few decades as robotic manipulators were slowly integrated into manufacturing, some forward-thinking inventors took advantage of improvements in computer technology to produce machines with programmed safety features, higher precision, and more versatility.

FLEXIBLE AUTOMATION

In 1947, Del Harder, an executive at Ford used the term "automation" to describe the increased use of "electro-mechanical, hydraulic, and pneumatic special-purpose production and parts-handling machinery." Harder was referring to "fixed" automation, in which machines are dedicated to one particular task. The new goal was to design programmable robotic manipulators to perform multiple functions. An early attempt at the new "flexible" automation was a patent for a dual-arm, gantry-mounted hydraulic machine with detachable grippers, filed in Britain by Cyril W. Kenward in 1954 and published in 1957. Kenward's manipulator was in some ways ahead of modern hydraulic robots because its hydraulic system and wiring were internal to the manipulator, rather than set in a separate base. However, control technology was not yet on par with Kenward's written design; and he was unsuccessful in promoting his patent.

By 1959, researchers at the MIT Servomechanisms Lab were demonstrating computer-assisted manufacturing; and a year later the American Machine and Foundry (AMF) Thermatool Corporation shipped its first Versatile Transfer Machine, or VERSATRAN, a programmable cylindrical robotic arm designed by Harry Johnson and Veljko Milenkovic. Around the same time, the Planet Company had commercialized their Universal Transfer Device (UTD), marketed as PLANETBOT. The hydraulically powered *polar coordinate* arm had five axes of motion allowing for 25 individual movements. It impressed spectators by passing a baton through obstacles at the 1957 St. Erik International Trade Fair on Automation in Stockholm, Sweden. The company claimed its robot could be reset in minutes to perform an entirely different set of operations. Planet sold eight units of this first model, and among its first customers was a division of General Motors, which used it to handle hot castings in a radiator production operation. Although by the 1980s, redesigned, hydraulic PLANETBOTs were being successfully utilized in forging operations, the first model's cumbersome mechanical-analog computer did not live up to expectations; and the machine behaved erratically when the hydraulic fluid was cool.

UNIMATE

The success of flexible automation has been attributed to self-taught inventor George Devol, Jr. On December 10, 1954, he filed a patent for a track-mounted, polar coordinate robotic arm, "Programmed Article Transfer." This was a general-purpose robot that could be adapted to a variety of tasks. His design included an electronic feedback controller, and a *teach pendant* (that he had invented earlier) for programming instructions stored on a magnetic drum. Magnetic read heads connected to the moving parts of the arm and base would scan a magnetically coated surface comprising wafers of ferrous material. This system provided position feedback information, controlling the rate of the arm's motion, and slowing it down before it stopped—a big improvement over the early PLANETBOT, that had only two speeds: off and full speed. Although engineers abandoned the ferrous wafers in later manipulators, Devol's basic concept of electronic programming became the standard for robot control.

Devol's plan was to sell the concept of flexible or universal automation that he called, "unimation." Machines that could do a variety of tasks in sequence, like picking up an object from one place, performing some operation on it, and moving it to another location, would be far more useful than those that could do only one job. In 1956, when Planet and AMF were getting their robots off the ground commercially, Devol was still shopping around his design with little success. Then he met physicist and engineer Joseph Engelberger at a cocktail party; and they discovered they shared both a love of science fiction and an interest in developing robots. Devol, who went on to receive 40 robotics-related patents, was finally granted U.S. #2,988,237, *Programmed Article Transfer* in 1961. He sold the patent to Consolidated Diesel Corporation (Condec) in Danbury, Connecticut, which was renamed Unimation. Engelberger served as president of the company and chief promoter of the most successful robot line in the United States. Although Devol's patent design was for a gantry-mounted machine, the first UNIMATE was stationary, with the control mechanism built into the base. It used transistors instead of vacuum tubes; and could do more than one operation in sequence. In 1962, Unimation introduced the 3,500-pound MARK II, which was originally used to automate the manufacture of TV picture tubes. Between the mid-1960s and 1984 when Westinghouse purchased Unimation for $107 million, the company sold approximately 8,500 UNIMATES with about 60 percent of these used in the auto industry. Westinghouse did not do as well with the UNIMATE as expected and sold it to the French/Swiss company, Stäubli, which has expanded and improved the line.

In 2004, the professional association Institute of Electrical and Electronics Engineers (IEEE) awarded Engelberger their Robotics and Automation Lifetime Achievement award for establishing and advancing the field of robotics and automation worldwide. However, the "father of robotics" recalls that in the mid-twentieth century, U.S. industrial manufacturers were not so enthusiastic about his vision. The year he founded Unimation, he had a tough time convincing American industry that robots could do the job he claimed they could and still save them money. In 1962, he convinced a hesitant General Motors to take a chance with his company, and they installed a UNIMATE in their Ternstedt, New Jersey plant to sequence and stack hot, die-cast metal components. The automated die-casting mold deposited red-hot car parts like door handles into cooling vats from a gripper attached to its steel armature. Workers on the line would then trim and buff the cooled parts. Still, GM executives did little publicity about the robot, since they considered it experimental technology and some people in the company were uncertain about its long-term success. Although it took a few years for the Ternstedt UNIMATE to show a profit, it operated successfully on the line for years. When the Ternstedt plant closed it doors a few years ago, the Smithsonian Institution transferred the pioneering UNIMATE to its industrial history wing.

In the 1960s, American companies were more concerned with making quick profits, and could not get beyond thinking of robots as a passing, entertaining fad. Even years after robots were in use, Engelberger appeared on the *Tonight Show with Johnny Carson*, where he demonstrated one of Unimation's new models serving drinks. Soon he received calls from company representatives who wanted to rent the robot as an entertainment device. In contrast, when Japanese industrialists invited him to speak in their country, manufacturers immediately recognized the potential of robotics in all phases of Japanese production and service industries. They had already seen demonstrations of the AMF VERSATRAN, which had been shipped to the country in 1967; and by 1968 the forward-thinking Kawasaki Industries had begun production of hydraulic robots for the Asian market under a license agreement with Unimation. The Japanese embraced robotics and soon became the major user of robots in the world.

COMPUTERIZED ROBOTS

The invention of the integrated circuit in 1958 reduced the size of computers. The smaller size and increased speed was attractive to engineers working on improving robot controllers. By the middle of the 1960s, robot research was alive within the university system. It was not only the desire to computerize robots that inspired advances in the field, but the enthusiasm of a new generation of students and professors eager to work in the latest subfields of engineering and computing. While computer departments had grown out of the mathematics schools, most universities were not yet equipped with dedicated robotics departments. Students interested in working on robot projects had to find advisors, funding, lab space, and tools from different departments; and research was often a result of combining resources. Much support for robotics came from mathematics and computer science departments where a few of the founders of the infant field of artificial intelligence (AI) had recently established research labs. For instance, Carnegie Mellon, whose Computation Center was founded in 1956 by Herbert Simon, created an interdisciplinary program, the Systems and Communication Science Program in 1961. This program allowed cross-fertilization between those studying computer science, math, psychology, and business and electrical engineering. In 1959, Marvin Minsky and John McCarthy founded the AI lab at MIT, where computer-assisted manufacturing was a hot new research focus. In 1965, the Computer Science Department was created at Stanford University, and McCarthy, who had moved from MIT to Stanford, established an AI lab there. Some support came from government and manufacturing companies eager to improve automated equipment. For instance, an early computer-controlled revolute arm was developed in the Digital Systems Laboratory at Case Western Reserve University, in Cleveland, Ohio, with support from the Space Nuclear Propulsion Office.

Among the early prototypes for a computer-controlled robot arm was the air-powered ORM (the Norwegian word for "snake"), developed in 1965 by students Victor Scheinman and Larry Leifer at Stanford University. With 28 inflatable sacks sandwiched between seven metal disks, the Orm moved by selectively inflating groups of the sacks. Unfortunately, the project was abandoned because the arm's movements could not be repeated accurately. Marvin Minsky designed a serpentine-type arm with 12 joints controlled by a PDP-6 computer at MIT in 1968. His wall-mounted, hydraulically powered TENTACLE ARM could reach around obstacles and could lift the weight of a person. As I will show in Chapter 5, this configuration—essentially an interpretation of the human spine—has been adapted successfully in recent years.

By the 1960s, manipulator technology was also being adapted to the design of prosthetics. A year after the UNIMATE came on the market, the first robotic prosthesis was introduced. The RANCHO ARM, developed at Rancho Los Amigos Hospital in Downey, California, had six joints, which gave it the flexibility of a human arm. In 1963, it was acquired by Stanford University, where it was developed into one of the first artificial arms to be controlled by computer.

In 1969, Victor Scheinman developed the first successful electrically powered computer-controlled industrial robot as his Masters thesis project. In contrast to heavy, hydraulic, single-use machines, his STANFORD ARM, was lightweight, electric, multiprogrammable, and could follow random trajectories instead of fixed ones. Scheinman showed that it was possible to build a machine that could be as versatile as it was autonomous. From this time forward, robots would be able to perform complex movements, like those necessary for arc welding and assembly. Interest in robotics was spreading beyond the United States and Japan. Sweden embraced robotics early. The Swedish company Svenska Mettalverken was the first European company to purchase a UNIMATE. Shortly thereafter, the Swedish engineer Roland Kaufeldt built a robot that was used in the plastics industry.

1970s

Computer controllers figured heavily in the popularity of robots in industry, and this created wider demand for companies to produce them. Consequently, by the 1970s, Unimation began to face strong competition. For instance, in 1973, THE TOMORROW TOOL (T3), designed by Richard Hohn for Cincinnati Milacron Corporation, became the first commercially available minicomputer-controlled robotic arm. In Stockholm, Sweden in 1974, ASEA unveiled its IRB-6, controlled by the new Intel 4004 microchip (introduced in 1971); and the AI department at Edinburgh University demonstrated their prototype Freddy II autonomously picking and assembling objects from a heap of parts. A year later, David Silver also demonstrated the advantage of improved sensors combined with microprocessor control at MIT, where he unveiled the SILVER ARM, a small parts assembly manipulator that reacted to feedback from touch and pressure sensors.

In a 1974 experiment, the STANFORD ARM assembled a Ford Model T water pump, guiding itself with optical and contact sensors. That year, Scheinman formed Vicarm Inc. to commercialize the STANFORD ARM design using minicomputer control, and began demonstrating it wherever he could find an audience. One story has it that Scheinman was demonstrating his small robot arm outside a convention center, when he met Engelberger, whose now successful Unimation had a display. Engelberger was impressed, presumably as much with Scheinman's tenacity as with the robot prototype, and invited him to share some space in the Unimation booth. In 1976, he incorporated a microcomputer into the VICARM. However, at the time Scheinman was more interested in research than sales; so in 1977 he sold the Vicarm rights to Unimation, who marketed it under the name Programmable Universal Machine for Assembly (PUMA).

GM had done a field analysis and concluded that 90 percent of all parts handled during assembly weighed five pounds or less. With Mitch Weiss, another recent engineering graduate as lead applications engineer for PUMA projects at Unimation, PUMA was adapted to GM specifications for a small parts handling line robot that maintained the same *space intrusion* (taking up the same space while in movement) of a human operator. The new PUMA (which other seasoned engineers at GM had not wanted to bother with) became a successful paradigm, and is still used for learning in university labs.

Although by 1980, Unimation had become the dominant robot manufacturer with 3,000 UNIMATES in the field; America lagged behind Japan in its acceptance of robots for decades. While Japanese engineers established their first robotics association in 1971, it was not until 1975 that the Robotics Industries Association (RIA) was founded in Ann Arbor, Michigan. RIA is still the only trade group in North America organized specifically to serve the robotics industry. One reason was that there were so few robot manufacturers and users. In his book *Robots in Practice* (1980) Engelberger credited Japan with at least nine industrial robot manufacturing companies in sequence production that provided full application engineering and field support; and only four in the United States (Engelberger, 277). Another reason is a difference in worldview. While the Japanese were ready to invest in robotics for the long term, American investors could not see beyond quarterly stock profit reports to make such an investment. Secondly, Japan already had a labor shortage and needed to invest in a long-term plan to offset its manufacturing, service, and construction requirements. As I will show in Part III, this is the reason Japanese roboticists give for investing so heavily in all sorts of robots, including humanoids. Among their accomplishments was the DIRECT DRIVE ARM, developed by Takeo Kanade in 1981. Unlike earlier manipulators that used drive chains to transfer power from the motor to the robot's joints, Kanade designed his robot with motors inside each of the joints. This arm proved to be a model for subsequent commercial robotic arms, for it solved the problem of backlash and made robotic arm movements quicker and more precise.

1980s

In 1980 seven specialists including Victor Scheinman founded Automatix Inc. Taking advantage of improvements in sensor technology, Automatix became the first company to market industrial robots with machine vision. By the end of the 1980s, their robot was controlled by an Apple Macintosh II. Industrial robotics expanded both numerically and geographically over the next 20 years. Members of the International Federation of Robotics (IFR) include companies based in Germany, Sweden, Japan, France, Korea, the Netherlands, Italy, and the United States. Among them is the Swedish company, ABB who in 2002 was the first to sell 100,000 robots.

The French-based Stäubli Group (f. 1892) diversified into robotics in 1982 when it reached a licensing agreement with Unimation to distribute UNIMATES in Europe. In 1988, Stäubli purchased Unimation from Westinghouse, and established Staubli-Unimation, a company that now produces over 1,400 articulated robots a year including its new RX line of precision manipulators. Another robot manufacturer that emerged during that time is Adept Technology. Established in 1983, today the company boasts that it is the largest United States-based industrial robot manufacturers. One of the reasons for Adept's success is diversity. The company designs, manufactures, and distributes everything from complete manipulators to motion controllers, to machine vision technology. Their small parts assembly robots are used worldwide in such areas as automotive, medical, consumer product manufacture, and machine tooling as well as in the food and pharmaceutical industries. Some of Japan's most successful robotics-manufacturing companies are IRF members. Among them is FANUC Ltd., whose 22-year-old American division is a leading supplier of industrial robots worldwide.

Industrial robots are used in die casting, arc and spot welding, spray painting, glass manufacturing, plastic molding, forging, boring, pressing, and more. With the advent of small parts assembly manipulators, the use of robotics was extended further into automobile finishing, consumer electronics, and food packaging.

WHAT IS A ROBOT?

Despite differences in sophistication, all industrial robots comprise the same essential elements. The central feature is the *manipulator* or arm—the part of the machine that handles objects. Some manipulators are configured with a number of joints, generally analogous to the human shoulder, elbow, and wrist. Robots are distinguished from other mechanized industrial machines by the dexterity that the manipulator gains from joint movement and from interchangeable *endeffectors*: grippers, scoops, brushes, spray-painting nozzles, spot- or arc-welding guns. Recent endeffector innovations like artificial hands and surgical tools have added to the value of using robots because they can perform precision maneuvers in dangerous environments such as space or contaminated sites; or provide precision placement and stable dexterity in exacting tasks like microsurgery.

Manipulator joints move in a variety of directions and at different angles referred to as *degrees of freedom (DOF)*. The joints of three degrees of freedom (3DOF) arms move along the X, Y, and Z (horizontal, vertical, and depth) axes only. A 6DOF manipulator also performs pitch, yaw, and roll movements that are measured as angles. Pitch is an up and down movement like opening a box lid. Yaw is left and right movement similar to opening and closing a door on hinges, and roll is rotation. Robot manipulators are classified according to their primal joint configuration. The inner or primal joints provide the gross positioning, while orientation is provided by the outer or distal joints. Robot arm geometry expresses the boundaries of the robot's workspace or work envelope, and is described according to the movement of the joints in space. Four types of arm geometry are Cartesian (or rectangular), revolute, cylindrical, and spherical.

The robot must have some power source, perhaps hydraulic, pneumatic, or electric. The power supply is the type of energy used to start the manipulator and keep it going throughout the operation. A power source is also needed to run the *actuators*. Actuators are used to drive the joints of the manipulator (although all joints may not be powered). System actuators use hydraulic, electronic, and pneumatic signals to convert energy into motion. The advantage of pneumatic actuators is that the compressed air is readily available on shop floors; however, with a maximum pressure of about 100 psi, they are incomparable to a hydraulic actuator, which can typically achieve a pressure of up to 3000 psi. Actuators include electrical motors, pistons, and today electroactive polymers, and other conductive materials. Important factors for actuators include the ability for precision control, operating life, and the ratio of power consumption to work achieved. Each actuator usually causes rotary or linear motion along an axis.

The second feature is the *controller*, the means by which the manipulator communicates with its processor or with the operator. The first robots were mechanical. These nonservo robots used a system of mechanical stops and/or limit switches to control their axis movement, rather than feedback from position sensors. Servo robots are controlled via sensors that continually monitor their axes for positional and velocity feedback information. This feedback is different from the pretaught or programmed information stored in a computerized robot's memory. Servos control the robot's movement on either a *point-to-point* basis, or *continuously* along its entire path or program of actions. Beginning with the UNIMATE, robot control was computerized; and advances in computer control resulted in more effective and versatile robots.

PATH SEQUENCING

The idea of teaching machines to do their jobs was significant to the growth of robotics. In order to be completely automatic, they must know what they are expected to do once the operator is no longer in the picture. Even the earliest robotic systems offered this capability. Before George Devol invented the UNIMATE, he adapted a lathe machine to learn sequences of operations. After connecting a magnetic recording device to the lathe,

> We turned out whatever parts we wanted, and in the process of making them, we magnetically recorded all the lathe's actions. From that point on, the lathe could automatically produce identical parts. (Devol in Logsdon, 50)

Portable teach pendants, similar to the one developed by Devol, have been used to control robot movement. In this process, positional data points are generated. They are stored by recording the movement of the robot arm through a path of intended motions within a determined space.

The robot follows a particular path, which is described as point to point, continuous, or controlled. With point-to-point programming, the manipulator moves from one point to another within the workspace. Typically, movement from one point to the next is not in a straight line. The joint actuators operate independently to arrive at their new positions; thus orientation of objects held by the endeffector may vary. During controlled-path programming, the robot is moved along a computer-generated, predictable path from point to point to the end of a task sequence. The path may be a straight line with endeffector orientation or it may involve curved paths through successive points with gradual changes in orientation. The computer calculates the coordinate transformations that control precise movements along the path. Manipulators that move in a continuous path, perform a sequence of movements that were stored in the robot's memory during a teaching sequence. The robot then automatically and repeatedly performs the entire sequence.

THE LONG ARM OF ASIMOV'S LAWS

Since their first appearance in "Runaround" (March 1942), Isaac Asimov's "Three Laws of Robotics" have served as the matrix upon which the plots of dozens of robot tales have been built. Asimov's androids operate according to the principles that a robot:

> may not injure a human being or through inaction allow a human being to come to harm.
> must obey orders given it by a human except where such orders would conflict with the first law.
> must protect its own existence as long as such protection does not conflict with the first or second law. (Asimov, 30–55)

Knowing that people have a tendency to "blame their tools," he made his robots incapable of harming their human operators. For example, in "Runaround," a robot that is designed only to operate with a human on its back ignores this programming when one of the men is in harm's way. It must move to save him because the first law supersedes the program: A robot cannot "through inaction allow a human being to come to harm." Although the Three Laws were invented to deal with the humanoid robots, Asimov's inspiration came from real-life engineering. He and Joseph Campbell (then editor of *Astounding Science Fiction*), with whom he developed the idea, reasoned that automated machines would need to be programmed with fail-safes to prevent their human operators from harm.

DANGER! DANGER!

Engineers remind us that in real life, Asimov's Laws are irrelevant until robots can actually recognize human beings as something other than an obstacle in their path or a goal to reach. Yet something akin to the Three Laws might have come in handy in factories where the first two generations of robots were used to speed up manufacturing. A number of industrial accidents killed or maimed human beings, simply because of the lack of sensor and control mechanisms, and lack of human/robot interaction experience. In one case involving a UNIMATE, a worker who had entered the danger area unexpectedly was pinned between the robotic arm and a safety pole meant to stop the robot's rotation. A similar tragedy stuck at a Kawasaki plant in Japan. For safety, their large, lathe-loading robots were fenced in. The safety protocol was for workers to enter this restricted area only through a gate, which would automatically trip a control that shut down the robots. Since production slowed when anyone entered the restricted zone, the worker responsible for the efficiency of the operation decided to keep the robots running by climbing over the fence. When he entered the area, a robot impaled him and then crushed him to death. (Logsdon, 21) In both cases, the tragedies were a result of worker error. Anticipating these sorts of problems, Asimov reasoned that with adequate AI, sensors, and control protocols, the machines could prevent even an obstinate human worker from harm.

LIMITATIONS

Industrial robots are fixed in their workspace. They might be set on rails, gantries, attached to walls, or to bases on the floor. However, even as early as the 1940s, some researchers were looking for ways of extending the capabilities of robots beyond the shop floor. This required mobility, and in turn better sensing capabilities. Mobile robots (the subject of Chapter 5) are also the result of years of research in AI, a field in which disputes over the nature and method of simulating intelligence eventually led to more versatile robotic machines. The relationship of AI to the expansion of robotics is significant, and will be covered in more depth in Chapter 4.

4

Smarter Machines: Artificial Intelligence and Robotics

◆

There are now in the world machines that think, that learn, and create. Moreover, their ability to do these things is going to increase rapidly until—in a visible future—the range of problems they can handle will be coextensive with the range to which the human mind has been applied.

—Herbert Simon and Allen Newell (1958)

Until the middle of the twentieth century, a "computer" was a person who performed arithmetical calculations or solved mathematical problems. Once the designation "computers" was assigned to machines that did this sort of work, researchers would further extend the man–machine analogy by trying to mechanize other sorts of human thought processes. In the late 1950s, research began in earnest to provide computers with the ability to perform human activities normally thought to require intelligence. Broadly, these virtual activities have been games (like checkers or chess), language recognition, and diagnosis (of technical problems or illnesses). Programs and devices resulting from this research are referred to as *artificial* intelligence (AI). The story of AI is a long and complicated one, but it is useful to touch on a few of its highlights here, for it was at the intersection of competing theories of AI that new approaches to robotics emerged.

ALAN TURING

Like Charles Babbage before him, British mathematician Alan Turing suggested that machines could do more than crunch numbers; potentially they could analyze information and solve problems. In "On Computable Numbers" (1937), Turing introduced a conceptual model of a universal computer, now referred to as a Turing Machine. It comprises a computing unit and an infinitely long paper tape on which it can write and read binary code. In the section of his paper on the existence of unsolvable problems, Turing theorized that if a problem cannot be solved on a Turing Machine, it cannot be solved by any other machine; nor by a human. Thus, implicitly, a computer can solve any problem that a human can. Logician Alonzo Church, with whom Turing had worked at Princeton, had independently come to a similar conclusion. The analogy between human and mechanical problem solving in the "Church-Turing Thesis" reflected a growing interest in practical approaches to machine intelligence.

Later, during World War II, Turing was a key figure in the development of the Bombes, which located keys to the German encryption system ENIGMA. He also contributed to the birth of AI theory by producing models for theorem proving, game playing, and decision-making programs. Today however, he is remembered best for the theoretical model in which questioners would have to guess whether the respondent was a human or a machine by their answers. During the 1940s, Turing proposed a computer modeled on how he thought our brain acquires intelligence. His idea was to build into a machine the propensity (inclination) to learn—something Turing assumed was innate in humans. The result would be a teachable intelligent machine. To assess the degree to which the machine is learning, he proposed a test using a teletype machine that he later described in "Computing Machinery and Intelligence" (1950). If after a designated number of questions, a person could not determine whether they were conversing with a machine or a human, the machine could be deemed "intelligent." This method of assessment is what is known as the "Turing Test."

INNOVATIONS

A swift succession of computing innovations following the invention of the transistor in 1947 by William Bradford Shockley, Walter Hauser Brattain, and John Bardeen made it possible to give Turing's theory material form. The transistor is a miniature device made of semiconducting material that can switch currents on and off faster than the bulky vacuum tubes used in earlier computers. It made practicable the stored program computer earlier described by Turing, von Neumann, Mauchly, Eckert, and the postwar computing innovations described in Chapter 3. In 1949, stored-program prototypes were being built. The first were the Electric Delay Storage Automatic Computer (EDSAC) by Maurice Wilkes in Britain, and the Binary Automatic Computer (BINAC) by Eckert and Mauchly in the United States. Despite the programming and financial difficulties that beset those pioneers, IBM was soon to commercialize such computers. In the summer of 1949, Eckert and Mauchly also developed a new computer that could accept plug-in peripheral devices, the Universal Automatic Computer (UNIVAC). The first unit was delivered to the U.S. Census Bureau in 1951. By 1955, IBM had introduced a calculator whose 2,200 transistors provided the calculating power of 1,200 vacuum tubes.

That same year, Shockley founded the Semiconductor Lab, giving birth to Silicon Valley—Land of Oz of the computing industry. The integrated circuit, introduced in 1958 by Jack St. Clair Kilby of Texas Instruments, is the basis of the multitransistor computer chip independently developed by Kilby and Robert Noyce during the next year. In just over a decade, researchers had produced the technical foundation for the machine Babbage's partner Ada Lovelace had described a century earlier, in which "the theoretical and the practical are . . . brought into more intimate and effective connection with each other" (Woolley, 1999, 268).

In that decade after the end of the war, a number of researchers began testing the limits of Turing's Machine, and to experiment with various "fixes" to those limits. Among them was Claude Shannon at Bell Labs, who had earlier described a mathematical model for communication. Two young researchers, who had worked with him over the summer of 1953, were experimenting with computer models of mind. They were John McCarthy, who had since earned a doctorate from Princeton and was teaching at Dartmouth College in New Hampshire, and Marvin Minsky, a fellow Princeton graduate with a Harvard fellowship. Another enthusiast was Nathaniel Rochester, who had just designed the IBM 701, the first general-purpose, mass-produced electronic computer. Together they would organize an interdisciplinary meeting to articulate the goals of a new field of research.

DARTMOUTH

In August 1955, McCarthy, Minsky, Rochester, and Shannon proposed the first interdisciplinary conference on what McCarthy termed "Artificial

Intelligence." The theoretical basis for the event was the idea that "every aspect of learning or any other feature of intelligence can in principle be so precisely described that a machine can be made to simulate it" (AI Conference Proposal, 1955). This was not exactly a new idea. After all, the discussions at the Macy conferences on cybernetics during the 1940s involved the same sorts of discussions. Walter Pitts, who with Warren McCulloch devised the first theoretical model of a neural net, recalled that the Macy meetings often began with the question of whether a machine could be built that would do a particular thing. Ultimately, they agreed that as long as one specified exactly what one wanted the machine to do, in principle, a machine could do it.

The two-month long Dartmouth conference, to be hosted by McCarthy and underwritten by the Rockefeller Foundation, was meant to produce a coherent trajectory for intelligent machine technology. During the year leading up to the conference, Herbert Simon, Allen Newell, and Clifford Shaw were all working on a simulated air defense station for the Systems Research Project at the Rand think tank in California. At the time of the conference, Simon was a professor of industrial administration at Carnegie Technical Institute (now Carnegie Mellon University) the department where Newell had founded a graduate program after the war. Newell and Shaw, a systems programmer, had developed simulations of radar maps for air defense exercises using a card-programmed calculator that could generate points on a two-dimensional map. Simon was impressed with the notion that computers could manipulate symbols, and began working with Newell to produce such a program.

Their initial goal was to write a chess-playing program; but eventually they produced a system that could search out and discover proofs for geometry theorems using symbolic language. Simon developed the *heuristics*—or rules of thumb that humans use to solve problems, while Newell and Shaw wrote an information processing language (IPL) that could mimic human memory processes. By December 15, Simon had successfully produced a paper version of the first proof. That summer, they demonstrated the result of their efforts, LOGIC THEORIST, at the Dartmouth conference. The program, which used recursive search techniques to simulate the logical properties of human intelligence, could prove theorems given in Bertrand Russell and Alfred North Whitehead's famous *Principia Mathematica* (1913).

Only a handful of researchers (mostly mathematicians) joined the organizers at Dartmouth that summer. Among them was Oliver Selfridge, who had helped edit *Cybernetics* for Norbert Wiener, and who had just developed a pattern recognition program called PANDEMONIUM at MIT's Lincoln Laboratory. Not all computer scientists accepted the hypothesis that machines could be intelligent, and many were simply not initially interested. Still those in attendance speculated that by around 1970 a digital computer would become a chess grandmaster, would uncover new and significant mathematical theorems, compose classical music, produce acceptable translations of language, and understand spoken language. Although no common conclusions or plans emerged from Dartmouth, specialists in a number of fields soon recognized the potential for intelligent machines—from business and industrial applications to studying neurological problems through virtual models of the brain—and AI was firmly established as a multidisciplinary science. Early AI research activity centered mainly (though not exclusively) around Carnegie Tech, MIT, Stanford, and IBM.

In 1959, McCarthy and Minsky founded the AI Lab at MIT, where researchers in the Servomechanisms Lab were already demonstrating computer-assisted manufacturing. After a few years, McCarthy moved on to establish an AI lab at Stanford University in California. AI labs were established in the computer science departments in many other major institutions. Many of these labs would also become the center of robotics research. The military, which had come to recognize importance of computing for launching and controlling rockets, began major funding for computer research in 1958 under its new Defense Advanced Research Projects Agency (DARPA). By 1960 DARPA was making significant increases in funding for computer research, in part because of the promise of AI. Computing quickly moved beyond single-purpose devices to general-purpose thinking and learning machines, and researchers produced a number of high-level programming languages.

The first problem in building intelligent machines is to define intelligence. At mid-century, researchers were discovering the limits of the Turing Test to do so. The second problem was an old one: How exactly does the brain process knowledge? The variety of approaches to AI that have emerged over the years grew out of two basic theories of Mind: Symbolic AI (also called Strong AI), described as a computer model of Mind, and connectionism, called the brain model of Mind.

SYMBOLIC AI

Symbolic or "Good Old Fashioned Artificial Intelligence" (GOFAI) is characterized by formalism and statistical analysis. Its proponents think of the human brain as working like a computer. They assume that human intelligence is essentially the ability to process internal or mental representations or *symbols* of the things we encounter in the world, and that all such information can be expressed in declarative form (rules or facts). This is the foundation of rule-based production systems, which have three basic components: a global database, or "working memory" of information necessary for the computer to perform operations; a set of rules, or "long-term memory," which states the conditions under which certain operations are performed, and a "controller" or rule interpreter that determines which rule is to be activated next. Human input controls the breadth of the database as well as the controller's protocol for achieving a goal.

Allen Newell described thought as problem solving. Humans consider a set of finite options for their actions based on preset criteria for the most satisfying outcome. Rather than searching through every single option, according to Newell our minds settle on the first one we arrive at that satisfies our preset criteria. Not surprisingly, LOGIC THEORIST, the program he developed with Simon and Shaw, used symbolic representations of real-world items and experiences to solve problems in this way.

They improved upon LOGIC THEORIST with their program, General Problem Solver (GPS) (1957), which used what they dubbed "means-ends analysis": GPS could analyze various aspects of a problem and reduce the difference between the current state of affairs and the goal state in order to perform a task. Simon and Newell were influenced by studies being done in psychology about how people puzzle out tasks. As experts in industrial administration, they could draw on their experience with the way organizations work to relate to these studies. In organizations, workers perform individual tasks that get the organization closer to its overall goal, despite the fact that the individual may have no knowledge of or direct influence on the goal. Newell had also been impressed with Selfridge's idea that complex goals could be achieved through the interaction of a number of subprocesses. Simon and Newell wrote GPS as a series of tasks and subgoals. For instance, where the goal is for a (virtual) monkey to fetch a banana hanging from a high place in a room where the only other object is a chair, the predefined tasks would be: "move chair; climb onto chair; move self; reach; grasp banana." The programmers had to provide preconditions for this to work. For example, the monkey could not "climb chair" until it performed "move chair." In addition, GPS had to have a set of location coordinates in order to make the ordered movements achieve the goal of fetching the banana. All this information, called a "difference table," is what GPS used to reach the goal. It then used means–ends analysis (the reduction in the difference between current and goal states). The program would identify the subgoals necessary to reduce the difference to zero, for instance: "get to chair." It could also apply backtracking ("If something doesn't work, try something else"). Programs that provide this sort of latitude could theoretically be the basis for the flexible control of automatic machines like robots.

A high-level symbolic list processing language—LISP—invented by John McCarthy at MIT in 1958, allowed for more efficient searches. LISP was able to deduce which of the items on its list were more suitable to solving the problem at hand, and could manipulate its own list to solve problems more efficiently. Like LOGIC THEORIST and GPS, LISP used a technique called recursion. In recursion, the program repeatedly calls upon itself to find all the possible solutions to a problem, and then reduces the list to the best possible solution. If this seems like a case of determining the best odds, it will not be surprising that recursion is used in game-plying programs like DEEP BLUE, that in 1997 beat world chess champion Gary Kasparov. (Critics note, however, that by 1997 DEEP BLUE supplemented recursion with databases of successful moves from most of the Grand Master games of the twentieth century, and that it had the advantage of powerful microprocessors not available in the early days.) As impressive as LISP programs were, their drawbacks were the immense amounts of memory and time required to sort and modify the long lists.

SHRDLU and Language

Other researchers explored the idea of helping computers to solve practical problems by creating internal or virtual worlds. Given the limited processing power of the time, programmers created equally limited environments using symbolic representations to teach machines about language. For instance, in the late 1960s, Terry Winograd, who eventually became a professor at Stanford University, did his doctoral thesis at MIT using a virtual world of blocks. The "micro block world" was a paradigm-in-progress, begun in the early 1960s by Minsky and a number of his students and colleagues. Among the important interim achievements was edge recognition: the ability of the computer to understand where one object ended, and another began. The earlier projects involved hooking a camera up to the computer so that it could "see" objects, and then learn to describe them, distinguish between them, and manipulate them. Winograd's system, called SHRDLU, was entirely virtual. The computer had no real vision; the monitor was there as a display for the benefit of the operator. Winograd's virtual world contained blocks of varying shapes, sizes, and colors, and a virtual robot arm that could manipulate the blocks the same way that a real robot manipulator would. The movements were displayed as line drawings on a computer monitor. (Computers of the 1960s did not have the processing power to rapidly render and animate 3-D imagery—something we now take for granted.)

SHRDLU was not one program but three: grammar, semantics, and deduction. This was a heterarchical rather than a hierarchical combination of programs. That is, the responsibility for control of the system can be distributed so that at any given time one or another part is in control depending on the need. The dialogue between the human operator and the program appeared on the screen in the form of subtitles. The operator would type in a request, for example, "Place the small blue block on top of the large red block." The computer would respond onscreen, "OK," and a visual of the movement would appear onscreen. If the operator asked a question like, "Where is the big red block?" The program's answer would appear on screen: "The big red block is on the table beneath the small blue block." If the request were ambiguous, the computer would let the operator know. For instance, in a virtual world that contained three different pyramid blocks, the request, "Pick up the pyramid" would elicit a response such as, "I don't understand which pyramid you mean." SHRDLU also had to know that if an object was in the way of the movement it was asked to make, it would have to clear the obstacle first.

This is all far more difficult for a computer than we might imagine. If the operator types, "Now pick up a bigger block," the computer would not automatically assume the operator was referring to the block the computer had picked up immediately before. It must figure out "bigger than *what*?" Words like "it" were ambiguous, since the computer could not *really* see the blocks. The operator would have to differentiate which "it" though specific language. The learning that took place in projects like SHRDLU was reciprocal, because it helped researchers recognize all the subtleties of language they would have to deal with in order to make computers useful, either as information machines or as controllers for other machines like robots. Winograd had begun his project with the assumption that one cannot separate linguistic knowledge from practical knowledge. It would not be enough for a machine to see objects, but to associate words and concepts with the objects. Yet in reality, SHRDLU understood nothing about real boxes, or its virtual arm. It could not have passed the Turing Test.

Although Symbolic AI would be the standard for programming for decades, the field occasionally experienced an enthusiasm slump referred to in the industry as an *AI Winter*. The media would be brutal in its reminders of the early promises of AI proponents, and worse, funding might be cut for research and development for a time. The one advantage to these declines was that it forced researchers to think more creatively. A good example of this is the growth of expert systems. Frustrated with the unwieldy universal programs, some researchers focused instead on building computing systems that could solve problems by manipulating specialized information. The combined advantages of parallel processing, more powerful computer chips, and "focused" programs were attractive to many organizations, who could immediately see their practical and economic advantages.

Expert Systems

The first of the "expert systems"—enormous global data bases devoted to very specific kinds of technical knowledge—was DENDRAL, which was used to analyze chemical compounds. Edward A. Feigenbaum, a former student of Herbert Simon, began the project in 1965. Although it took 10 years for Feigenbaum and his colleagues at Stanford to complete the program, DENDRAL demonstrated the value of such systems, and inspired a knowledge-engineering industry that was generating billions of dollars in revenue by the late 1990s. Among these systems was MYCIN, which helped physicians diagnose and find appropriate remedies for specific illnesses. MYCIN, developed in 1975 by Stanford doctoral student Edward Shortliffe, could diagnose infectious blood diseases like meningitis. In 1979, an article in the *Journal of the American Medial Association* (*JAMA*) reported that in a comparative study MYCIN's diagnoses were equal to those of human medical experts. In the early 1980s, another expert system that received a lot of press was PROSPECTOR, a system that extrapolated information from geographical surveys to identify promising mining sites. The system's rules list was described as containing the collected knowledge of nine geologists.

Around the same time, Diesel Electric Locomotive Troubleshooting Aid (DELTA) made good on its promise to reduce the workload of an engineer at General Electric. The DELTA story demonstrates the logic for replacing human specialists with AI: General Electric employee David Smith was the only one competent enough to troubleshoot locomotive engine problems satisfactorily at the time, so he routinely was sent to remote locales. This cost a lot of time and money and put a good deal of stress on Smith, who could only be in one place at a time. However, by encoding all the same diagnostic knowledge and decision-making ability as Smith, his expertise could be used in many locations simultaneously. By 1984, DELTA was diagnosing 80 percent of the company's locomotive breakdowns with the expert information and experience shared by Smith, and providing detailed visual repair instructions via a videodisc playback unit.

The Brittleness Bottleneck

The idea continued to circulate that if a Turing machine could fool a person conversing with it into believing that he/she was conversing with another human being, the machine could be judged intelligent. However, it would be a long time before a computer could fool a person in this way. For instance, computers that observe the outside world and use the information to determine how to repeat what they see were idiosyncratic, in part because vision systems were not as sophisticated as they are now, but especially because they lacked what we refer to as common sense. Because of their inability to deal with the ambiguity and subtleties of linguistics, these systems were considered inflexible or "brittle." For example, an explanation-based system (EBS) can mimic a demonstration of stacking blocks, but it does not really understand about gravity—that blocks can only be stacked from the ground up. It does not understand the symbols; it only manipulates them according to rules.

A number of researchers have tried to explain how much of our understanding about language we take for granted. Minsky once offered this example: a computer could be told that birds can fly; but would also need to know the many exceptions: ". . .unless they are penguins or ostriches; or if they happen to be dead, or have broken wings, or are confined to cages, or have their feet stuck in cement, or have undergone experiences so dreadful as to render them psychologically incapable of flight" (Minsky in Kurzweil, 93–94).

In a lecture given in 1998, Hubert Dreyfus provided his own explanation of why it is so difficult for a machine to grasp the nuances of language. Dreyfus set up the following scenario:

> Today was Jack's birthday. Penny and Janet went to the store. They were going to get presents. Janet decided to get a kite. "Don't do that," said Penny, "Jack has a kite. He will make you take it back."

Human beings, Dreyfus explains, would take for granted that the presents are for Jack, and that the kite is meant to be a present for Jack. However, a computer program would not necessarily get this. One could program a "frame" that includes information about birthdays and gifts; but this AI version of Miss Manners would not solve the problem of the last "it" in the scenario. Grammatically, that word should refer back to the last mention of "kite," though a human reader would be able to guess that the "it" Jack would make Janet take back would not be his old kite, but the one Janet bought as a gift. However, a computer programmed with the laws of

English grammar would misunderstand and "assume" Jack "will make you take [the kite he already has] back."

Perhaps the best illustration of the language problem came not from a computer expert, but from a science fiction writer. In 1973, Rachel Cosgrove Payes published a humorous short story, "Grandma was Never like This," written as a letter of complaint to a service robot company. The correspondent describes several incidents involving humanoid babysitting robots that he and his wife have rented from the company. In the first incident, they give the robot explicit instructions that under no circumstances should it let anyone in the house. Consequently, they return home to discover that the robot has changed the locks, and will not even let *them* in. To solve the problem, they blow smoke through a crevice in the door and yell, "Fire, fire—save the baby!" As the robot—once more just following instruction—runs from the house with the baby cradled in its arms, the couple grab their offspring out of the robot's hands. The husband writes that the couple agreed to take on a new robot. This time, on the way out of the house, the mother shouts to the new robot, "Don't forget to change the baby!" The husband concludes his letter: "Please return our baby, as we don't want the one your robot changed ours for (Payes in Elwood, 55)." Payes's story anticipates Dreyfus's skepticism that we could ever create rules sufficient for a computer brain to understand the fluid ways that humans use language. However, one researcher disagrees, and he has been working to solve the problem for two decades.

Common Sense

Douglas B. Lenat, president of Cycorp Inc. has attempted to break the "brittleness bottleneck" with CYC®, a project he began thinking about while teaching computer science at Stanford. He began with the assumption that in order to operate in the world, people require a lot of common sense. Therefore, in order for an AI to be useful to people, it needs the same kind of basic knowledge. For instance, it would need enough information both about language and about taking care of babies to understand that when the mother in Payes's story asked the robot to "change" the baby, she was not asking it to *ex*change the baby.

Lenat's timing was very good, because in the mid-1980s a lot of leaders in the American high-tech industries had grown concerned about competition from the Japanese. They established a research consortium, the Microelectronics and Computer Technology Corporation (MCC) in Austin, Texas, and offered Lenat a job there. Aware that he would not be able to achieve his goal with university funding and student volunteers, in 1984 he joined MCC and began the project he named CYC® (for encyclopedia). At MCC operators spent years inputting millions of common sense facts, including information like, "Mothers are always older than their daughters; birds have feathers; when people die, they stay dead." After accumulating a certain level of expertise about the human world, CYC® began asking questions to clarify certain points. In 1994, Cycorp was spun off from MCC with government funding. As of 1997, over $40 million had been spent on the project, though Lenat was able to recoup some of the costs by marketing CYC® to corporations like IBM, Digital Equipment, and Glaxo Welcome.

One advantage CYC® has over more traditional database managers is that it can make choices based on analyses of information, rather than by simply matching key phrases. For example, inputting a photo retrieval request for images of "happy people" may yield a photo with the caption, "parents watching their daughter learn to walk." A request for an image of a "strong, adventurous person" might result in a retrieval of a picture captioned, "Man climbing a mountain." These are far more sophisticated responses than those of early projects like SHRDLU: CYC® has information about many things that make people happy. One of them is interaction with their own children. It also knows that a man is a person, and that mountain climbing is a dangerous, demanding sport, requiring a "strong, adventurous person." The idea is to use this kind of understanding to make other machines more flexible. Lenat realized that the more bits of information CYC® had stored, the better the chance that some of them would be contradictory. At one point, he partitioned the system to organize the data into contextual areas. Therefore, if CYC® is being questioned about vampires, it will know that Dracula is a vampire, but also that Dracula is not real.

Today CYC® is still a work in progress. Lenat has recently acknowledged that it may take another 20 years to perfect the system; however, the software is being used to improve the responses of internet search engines. Another method of solving the weaknesses of rule-based machines is to give them the ability to deal with ambiguity. This method is called "Fuzzy Logic."

Fuzzy Logic

"Fuzzy logic" is adapted from the work of logicians of the 1920s who concluded that measure is a matter of degree. Because it allows machines to make common-sense decisions rather than strict rule-based ones, it is useful when there is no algorithm or rule to follow for an ambiguous situation. For example, it is used in cases where a situation or an object cannot follow the law of the excluded middle (can't be 100 percent one thing or another, or 100 percent of one group and 0 percent another). Still, it must follow the basic law of formalism in that the sum of the degrees to which it is and is not must add up to 100 percent. For example, a washing machine can "decide" when the water changes from warm to hot. In *adaptive* fuzzy systems, the machine "learns" the rules by observing human regulation of devices—useful where there is no human expert to define the rules for shifting from one mode to another. Whereas probabilities (e.g., Bayesian theory) measure the odds that an event will occur or a decision will be the most effective, fuzzy logic just measures the degree to which something *may* occur or to which a condition *may* exist. The first digital fuzzy chips, developed by Masaki Togai at Bell Labs in 1985, processed 0.08 million fuzzy logical inferences per second.

Minsky once equated conventional symbolic production systems with filling out a tax form: adding up deductions and subtracting from liabilities. This is not a case of learning, but information management—calling upon the most appropriate choice under each condition. He has argued that symbolic AI such as expert systems will always be incapable of anything more than database searches because their knowledge of the world is limited to programmers' input. (Although more recently he has conceded that thought probably involves both rule-type reasoning and pattern recognition using parallelism.) An alternative to symbolic AI is connectionism, in which researchers attempt to create machine intelligence structurally, that is, by producing electronic analogies to the neural structure of the human brain.

CONNECTIONISM

Connectionism draws on research in a number of fields including computing, neuroscience, and behavioral psychology to offer learning-based alternatives to traditional symbolics that its proponents argue more closely resembles human brain function. It grew out of early discussions of cybernetics. During the 1940s, Warren McCulloch and mathematician Walter Pitts created a mathematical model of neural nets. Later research revealed some of their thinking to be faulty; nevertheless the concept was intriguing to many researchers. For instance, Minsky's first big project at MIT was a neural network that simulated a rat in the maze that he built with colleague Dean Edmonds from vacuum tubes and an old automatic pilot from the war. (His old teacher, Claude Shannon, had previously produced rats-in-a-maze models.) Minsky discovered that he could teach the minimally programmed computer rat to learn from its mistakes. In the mid-1950s, Nathaniel Rochester of IBM who designed the first mass-produced, electronic digital (binary) computer simulated neural nets on his IBM 704. The theory that AI should be designed to simulate neural activity is called *connectionism*.

Neural Nets

Human brain neurons collect electronic signals from others via dendrites and send out electronic spikes of activity along threadlike axons. At the end of each of the thousands of branches of these axons are synapses that convert activity from the axon into electrical effects that inhibit or excite activity in the connected neurons. What effect does this have on us? The degree of effect between neurons may be helping us to look, learn, or focus on a current state of reality; to move, to speak, or to wonder. Our synaptic activity is "weighted" in favor of any of these functions at a given time.

Connectionism draws from cybernetics the idea that thought is the result of sensory feedback between our central nervous system and the environment. Since it assumes that brain function is directly related to its structure, connectionists work with simplified models of the brain composed of a number of manufactured neurons. They apply weights (electronic markers that assign importance or value to some information) to simulate the strength of brain synapses or "connections" that link one neuron to another in our brain (hence the tag "connectionism"). These artificial structures are called artificial neural nets (ANNs).

ANNs can learn by repeated demonstrations of the correct solution to problems. Basically, an ANN breaks down complex problems into a series of individual ones, each assigned to separate "miniprocessors." Each of the individual operations communicates with the others through connectors. Neural nets require less programming than rule-based production systems because they can learn to achieve their goals through example, trial and error, and practice. Despite the slower processing time for individual neurons, overall response in such a "parallel processing" system is theoretically better suited to learning than a serial computer that devotes the problem to one central processor, and rules list programming. (Unlike early GPS that had to stop the main routine to run subroutines, parallel systems do the various tasks simultaneously.)

At mid-century, neural net theory was not widely embraced since researchers didn't really understand that much about our brain's neural connectors, and technology did not allow for modeling with more than a few ANNs. Ironically, Minsky (who had temporarily abandoned his neural net research around the time of the Dartmouth Conference) bears some of the blame for the 15-year stagnation of neural net research. He and MIT colleague Seymour Papert published a well-received critique of Frank Rosenblatt's "Perceptrons"—a more complex version of the McCullough-Pitts neural model—in which they complained that single layer nets were hopelessly limited. Today ANNs serve practical applications, for instance the "sniffer" that scans neutron reflections in airline baggage to detect plastic explosives. Neural nets are also being successfully used in face recognition, reading, and recognizing grammatical structure, as well as some of the robots that will be described in the following chapters.

Societies of Mind

In their attempt to model AI on the human brain, connectionists have tried to explain how it is that our brain gets and uses knowledge. In the mid-1970s, Minsky offered the concept of "Frames" to explain how we recall things, and then compare those recollections with immediate sensory experience. "When one encounters a new situation . . . one selects from memory a substantial structure called a frame . . . adapted to fit reality by changing details as necessary." This protocol includes a setting that Minsky called a "terminal," matching features with the image encountered, and comparing required properties or "markers" for each feature of the frame. For example, a wood floor is a marker for the frame "room." Encountering grass activates another frame—"outdoors (Minsky in Crevier, 172–173)." Minsky later described the enormous number of processors of human thought as "agents." Each agent draws connections between inputs (experiences) and "remembers" which sensory experience is good for each situation. Minsky called groups of them "agencies." Together, the agents form a "society of mind."

Herbert Simon thought that Minsky's frames were nothing more than the description lists that he and Allen Newell had built into IPLs in the mid-1950s. However, Minsky's inspiration was more likely the pattern recognition theory of Oliver Selfridge, with whom he worked at MIT in the late 1950s. According to Selfridge, the brain responds to sense experience not by retrieving data from an immense library of facts, but by calling on a small number of "experiences" to determine how to deal with the question or command at hand. Although Selfridge was not working with neural nets, his concept shares with connectionism both the idea of breaking problems up into separate ones, and calling on stored modules of experience rather than accessing individual facts.

Hubert Dreyfus had described his idea of "similarity recognition" in comparable terms: Dreyfus, a philosopher whose critique of symbol-based AI is discussed above, defined learning in humans as a five-step process from novice to expert, with novices working according to rules supplied from outside, and beginner, competent, and proficient learners using respectively fewer rules and depending instead on learned behavior. He thought that "experts" choose to act on previously successful behaviors (that is, recognition that something worked to solve a similar problem before) and not lists of rules.

Minsky later published an updated and expanded explanation of his frames concept in *Society of Mind* (1985–1986). At the same time, James McClelland's "Parallel Distributed Processing" papers demonstrated how by extending Rosenblatt's perceptrons to many layered structures, they could be made to overcome most of the failings Minsky and Papert had described in their 1967 critique. These concepts finally could be tested, since technology was catching up with theory: By the end of the 1980s, computer memory cost only a hundred-millionth of what it had in the 1950s. Five years later, fifth-generation computing machines were using massive parallel processing suggested by McClellan and his colleagues (the Connection Machine, by Thinking Machines Inc. had 250,000 processors), and by the mid-1990s, neural chips were on the market.

By the 1990s, developers of communications devices took advantage of theoretical and technological innovations to produce consumer products that were more user-friendly, theoretically because they "think" like we do. Engineer Jeff Hawkins articulated his design rationale in terms similar to the connectionist pioneers. After leaving engineering to study theoretical neuroscience, Hawkins concluded that the uniform structure of the Neo cortex implies that "the same basic mechanism underlies all sensory processing." He believes that brains are not like computers. His concept of "auto-associative memory" sounds quite similar to Minsky's frames. According to Hawkins, certain information is stored; but new sensory information taps into the stored memory, which is basic knowledge, and predicts what will happen. The more efficient a person is at predicting complex patterns (what ought to be happening next) based on certain visual, auditory, sensory input, the more "intelligent" a person is. For example, when encountering a face, we look for expected patterns: If we see an eye, we expect another eye to be next to it, a nose below it, and a mouth below that. When this is not what we find, we judge that something is wrong. He designed his Palm Pilot and Handspring on these principles.

The debate has continued between computer and brain models of Mind. Until he died, Newell continued to use symbolic representations to develop AI like the SOAR project he began in the 1980s that he referred to as a unified cognitive system. Meanwhile connectionists, referring back to cybernetics, argue that learning and problem solving is a product of the interaction of our entire nervous system, which is responding to environmental experience as a pattern of activity triggered by actual physical experiences, not maps of them. Other academicians reminded us that despite the fact that that AI had not produced the HAL of the film *2001: A Space Odyssey*, computers and AI had lightened human beings' workload. It is true that computers are only the sum of their input, but once the data is there, they can remember it indefinitely, copy it precisely, and display it in many places simultaneously. Any bookkeeper can attest to the human cost in time, stress, and eyestrain in looking for a 3¢ difference, the result of nothing more than a human input error. Furthermore, intuitive search engines make it possible for people to communicate and retrieve information instantly across space.

In the 1980s, AI was experiencing another slump. This new AI Winter was the subject of the 1984 meeting of the AAAI. The blame fell in varying degrees onto different shoulders: many LISP-type programs were so long that they became too unwieldy to be useful. The promise of neural nets had still not been fulfilled. To a significant degree, the media was to blame, for it had the propensity to write headlines that exaggerated the actual claims of researchers. Some of the criticism came from the scholarly community. For instance, John Searle, Professor of Philosophy at Berkeley, took GOFAI to task for promoting the impossible. Searle argued that it was impossible for a digital computer to become intelligent. Although some have interpreted Searle's argument as a holdover of Cartesian dualism, he was very articulate in putting forward arguments against symbolic representations ever being able to understand the world in the way that people do. The discussion at the 1984 AAAI conference undoubtedly was inspired in part by Searle's attack, for his lectures were published in *Minds, Brain, and Science* (1984) to much publicity. Consequently, researchers once again considered alternative ways of expressing machine intelligence, and of building useful machines.

NEW DIRECTIONS

By the end of the twentieth century, some younger researchers, drawing on work in the cognitive, behavioral, and neurological sciences, concluded that learning is really a hybrid of what traditional symbolic AI programming extrapolates into rule lists, the parallelism and patterns of neural nets, plus behavioral modification. A number of projects meant to demonstrate how organic creatures navigate the world provided the theoretical and physical structures for a new generation of practical robots.

In order for robots to provide a variety of useful services, they would have to operate autonomously, safely, and effectively outside structured factory environments, sometimes in places inhospitable to humans, like nuclear disaster sites and outer space. Cybernetics and connectionism, which uses the lower-level brain function for its model, influenced a number of engineers to take a "bottom up" approach. They developed behavior-based robots (BBR) that use minimal programming and learn from their sensorial experience. Others extended the technology of heavy industry robots to service robotics. In the following chapter, I will show how in the last decades of the twentieth century, technical improvements in such areas as programming, battery power, artificial intelligence, and materials science made it possible for researchers to choose from a variety of methodologies to build robots suitable for diverse environments and uses.

Part III

GROWTH OF THE FIELD

5

Getting Around: Perception, Locomotion, and Power

◆

... of their own motion they could wheel into the immortal gathering, and return to his house: a wonder to look at.

—Homer, *The Iliad*, XVIII

In just over 20 years, robot-assisted manufacturing had become a standard, with dozens of industrial manufacturers supplying robot arms to companies around the world. Suppliers/purchasers, the Japanese Industrial Robotics Association (JIRA) estimated that (excluding fixed-sequence and manual manipulators) at the end of 1986 there were 116,000 industrial robots in Japan, 25,000 in the United States, 12,400 in Germany, 5,270 in France, and 2,380 in Sweden. Despite the technical innovations that had contributed to the expansion of industrial robotics, the limitations of robot control had become apparent to those working in the field. They registered their concerns in engineering reports published during the first half of the 1980s.

Some critics asked how industrial robots could meet the demand for increased speed and load capabilities stimulated by improved software when the mechanical aspect of robots had not really changed since the 1950s. Others recalled the First International Symposium on Robotics in Chicago in 1970, where speakers characterized a second generation of intelligent, autonomous, mobile robots as "around the corner." Yet a decade later critics complained that promise had not been fulfilled, particularly with regard to robot vision systems. For instance, Unimation and Automatix engineers had developed their first robot vision systems in 1980, and Adept Technology had just begun incorporating vision systems into their products in 1983. Existing vision and other sensing technologies were still inadequate for mobile robots. However a number of factors stimulated increased investment in advanced technologies, including robotics.

By the early 1980s, fierce international business competition, shifting workforce demographics, nuclear plant and other environmental emergencies, political upheavals, and ambitions to exploit extraterrestrial and underwater environments motivated government agencies to support robot research for exploration, recognizance, search and rescue, and disaster cleanup. For example, in 1983 DARPA initiated its decade-long Strategic Computing Initiative (SCI). Some of its billion-dollar budget was earmarked for autonomous land vehicle projects (ALV) discussed below. At public hearings, representatives from industries as disparate as mining and health care described the value of robots to address rising labor costs and to assist and replace human workers in tedious or dangerous tasks. These activities motivated both veterans of university research labs and commercial robot manufacturers to establish the first service robot companies. Within a decade, mobile robots were delivering hospital meals, picking up garbage and delivering mail in corporate environments, guiding museum visitors, and cleaning up hazardous waste. Autonomous vehicles were being developed for combat situations and robotic rovers were being readied to explore the surface of Mars. However, this apparent overnight response to the demand for new robot platforms was in fact the product of a generation of research that began during the early days of cybernetics.

GREY WALTER'S TORTOISES

Neurophysiologist W. Grey Walter (1910–1978) carried out some of the earliest experiments with mobile robots in the 1940s. Walter was one of the founding members of the Ratio Club, an interdisciplinary, United Kingdom-based group who met to discuss different ways of looking at the brain, and to explore many of the same questions as those raised at the cybernetics meetings in the United States. During this period, he was based at the Burden Neurological Institute in Bristol, United Kingdom, doing research in electroencephalography (EEG), a method of measuring human brain waves originally developed by Hans Berger (1873–1941). Walter designed his own version of an EEG brain topography machine through which he observed theta brain waves (associated with light) and delta waves (associated with deep sleep). He was interested in learning more about animal nervous systems; but since EEG is noninvasive, he turned to building electromechanical models of animals so he could observe and tinker with their artificial nervous systems. He playfully classified his shoebox-sized creatures as *Machina speculatrix* because they exhibited exploratory, "speculative" behavior widely displayed in the animal kingdom.

With the assistance of his wife, Vivian, he began building the artificial animals using electromagnetic relays and vacuum tubes for switching elements, and recycled gears and electric motors for actuators. By 1948, they demonstrated the first of a series of small three-wheeled mobile robots that he called "tortoises." The front wheel had a drive motor and a motorized steering assembly that could rotate 360 degrees in one direction. Each had a simple "brain" composed of only two neurons: a light sensor that controlled the steering spindle and a touch sensor on the shell that caused both sensors to oscillate. A photosensitive resistor mounted on the shaft of the steering assembly ensured that it always faced in the direction of travel. The bump sensor "skirt" would close a switch if the machine hit an obstacle in any direction.

Under moderate light, the steering motor stopped scanning, and the tortoise would move toward the light. When the tortoise was approximately 15 centimeters from the light source, a relay kicked in and the scan/steering motor ran at double speed. This "flee mechanism" made the tortoise avert from the light. When a tortoise encountered an object, the bump sensor would make it vacillate between trying to move forward and trying to turn away, which resulted in it moving around the object. The recharging station inside the tortoise's "hutch" had a light in it, and normally it would turn away before entering. However, as its batteries ran down, its sensitivity to light decreased. In this case, before it "saw the light," it would move all the way into the case and thus connect with its recharger. The tortoises exhibited a variety of complex behaviors, for instance choosing between different light sources, and performing the equivalent of mating and territorialism. Another interesting display occurred if a tortoise approached a mirror. It would become attracted to the reflection of the light bulb mounted on its own steering mechanism that indicated when it was on. This switched off the steering motor and its light. No longer attracted, the steering mechanism would switch back on, along with its light, causing the attraction again.

The original tortoises were not mechanically reliable and had to be frequently adjusted. Walter's technician W. J. "Bunny" Warren built six new and improved tortoises, three of which were exhibited at the Festival of Britain in 1951. Walter discussed his theories in two articles in *Scientific*

American, "An Imitation of Life" (May 1950) and "A Machine That Learns" (August 1951), and later in his book, *The Living Brain* (1953). He emphasized that the tortoises did in fact display unpredictable or spontaneous behavior in response to even minute changes in electric current, terrain, or light level.

Highly influenced by behaviorist theory, and especially the experiments of the Russian physiologist, Ivan Pavlov, Walter set out to build machines that could acquire and display conditioned reflex. Pavlov had repeated an experiment in which a bell (a neutral or "conditioned" stimulus) rang each time the dog received food. After a time, the dog would salivate whenever it heard a bell, whether or not it received food. Walter fitted a second generation of tortoises that he referred to as *Machina docilis* with a learning circuit. He used a 3000 Hz tone as the neutral stimulus, and the action of kicking *docilis* as an unconditioned stimulus. The kick stimulated its bump sensor, causing the tortoise to flee; at which point the tone sounded. Eventually, it was conditioned to flee at the sound of the tone without the kick.

Walter was not alone in exploring the relationship between biological organisms and their environment. Inspired by work done earlier in the twentieth century, psychiatrist and fellow Ratio Club member W. Ross Ashby published "Design for a Brain" (1948) which he expanded to book length in 1952. Ashby described the agent (organism) and its environment as elements of an absolute system, each imposing a reciprocal influence. Ashby explained that biological organisms (and theoretically machines) are self-organizing, that is, constantly drawing stimulus from the environment, and changing in the direction of equilibrium or stability. (Recall the ideas of equilibrium or balance articulated by the ancient mechanics discussed in Chapter 1.) He wrote a mathematical description of this process and built a model homeostat, a series of four units that exhibited this reciprocal influence. Ashby's work was well received and discussed by computing pioneers like Simon and McCulloch. The stimulus response approach of researchers like Walter and Ashby also influenced a later generation of robotics engineers to build the biologically based robots discussed below.

GOING MOBILE

Even in the early days of industrial robotics, researchers envisioned robots that could be useful outside a structured factory environment. Meanwhile, some AI research focused on fitting computers with sensors to explore theories of cognition. Mobile robotics emerged at the intersection of these two initiatives. Researchers had to address several practical issues if they were going to achieve their goal of autonomous mobile AI.

The challenges to mobile robotics may be broadly organized into four areas:

- Locomotion (Movement)—By what means should the robot move from place to place?
- Perception (Sensing)—How will the robot recognize its path, goal, or obstacles?
- Control (Computation)—Where will the robot get its goal plan; how will that be communicated to its actuators and manipulators? How will it know when its task is complete or communicate problems it is having completing them?
- Power (Energy Source)—What is the most effective way to keep a robot up and running, especially if it is to operate autonomously?

Fixed-place industrial robots and mobile robots share many requirements, such as task planning or force sensing. The problem of uncertainty is more critical for mobile robots. For instance, they may encounter a variety of unexpected terrains, atmospheric conditions, and obstacles as they move toward their goal. They must be able to differentiate solids from liquids and gases, and to distinguish a person from a bush, and a floor from a wall, or a stairway.

The proposed working environment for the robot influences the engineer's decisions about its locomotive power and suspension, size and shape of the chassis, perception equipment, and method of planning. Safety is always an issue in robotics, but the assumption in mobile robotics is that the machine will either be working among people and other living creatures, or that it will be alone in inhospitable environments. In either case it must be able to negotiate space in a way that is safe for both biological entities and for the robot. By the mid-1960s, a number of university labs were experimenting with mobile robots. Wheels were the most logical choice, since wheels are a well-understood and relatively inexpensive method of providing efficient and stable locomotion over flat surfaces like the laboratory hallways where the robots were being developed. The long-range goal was to build robots that could operate without human intervention.

The Beast

One of the first experiments in robot autonomy was carried out at the Johns Hopkins University Applied Physics Lab in the mid-1960s, where researchers produced a canister-shaped wheeled robot controlled by dozens of transistors. The HOPKINS BEAST was able to navigate down the center of the white halls using sonar until its batteries ran low. It would then seek out contrasting black wall outlets with special photocell optics, and use a touch sensor on its recharging arm to plug itself in. After recharging, it would resume its rounds of the hallways. The BEAST's deliberate coordinated actions have been compared to the bacteria-hunting behaviors of large nucleated cells such as paramecia. However its sensing capability was not complex enough to allow it to recognize any outlet as a potential source of energy. It could only recognize the photocell-marked electric outlets. Like Walter's experiments, it demonstrated the value of sensors to locate targets, distinguish objects, and avoid obstacles.

Despite the significant advantage of transistors over vacuum tubes, computers were still too unwieldy (and too few); so for mobile robots to perform any complex behaviors, they had to be connected to the lab's mainframe computer using radio signals or a cable. Therefore, while subsequent robots did display mobility, they were remotely controlled and therefore technically not autonomous robots. This was the case with SHAKEY, funded by DARPA between 1966 and 1972 as part of an initiative to develop mobile recognizance computers for the military. Nils Nilsson built and programmed the robot under the direction of Charles A. Rosen at the Stanford Research Institute Artificial Intelligence Lab (SAIL).

SHAKEY

SHAKEY, so named for its vibrating movement, consisted of a TV camera and controller, a triangulating range finder, onboard logic unit, and collision avoidance detector, all mounted on a platform.

It moved via drive wheels, and turned on casters. Originally, SHAKEY was controlled via radio and video links by an SDS-940 computer with 64K 24-bit words of memory, using FORTRAN and LISP programming. It did its problem solving in QA3. Around 1969 this system was replaced with a DEC PDP-10 (later PDP-15) computer with 192K 36-bit words of memory, and STRIPS for problem solving. It relied on hierarchical programming for perception, world modeling, and acting. Low-level action routines controlled route planning and simple movement and turning, while intermediate-level programs linked the lower-level ones together to accomplish more robust task execution. High-level programming provided task planning and execution. The system saved the plans for future use.

SHAKEY is often described as the first mobile machine to use reasoning; however it lacked the spontaneity of Grey Walter's primitive tortoises.

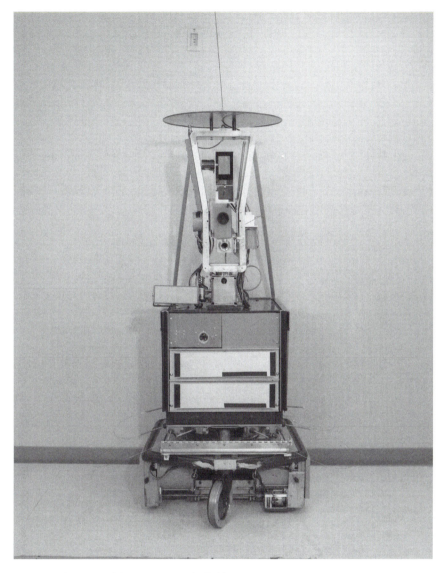

SHAKEY. Courtesy of SRI International, California

Before it even set out on its mission, it had to be programmed with a two-dimensional map of its goal area. Simple tasks took hours to complete because information was processed at speeds that are primitive in comparison to those of today's microprocessors. This required the robot to wait for sensor readings to be relayed to the mainframe, compared with its internal world model (IWM), which was updated and relayed back. Its primitive vision sensors could provide edge detection and proximity information, but the "real world" that SHAKEY was reasoning about had to be a very simple one where its objectives were simple geometric blocks, and not people or soft-edged inanimate objects.

The SHAKEY project established the IWM approach to controlling mobile robots. This may have been because researchers were used to programming industrial robots to move in a preset path, or because the programmers were devoted to the symbolic, or computer model of AI. The real value of the project is that it demonstrated that a robot could be designed to move to a goal and perform a task. It remained for researchers to improve robot reaction time and to give them autonomy. It would be years before robotics achieved this goal. Meanwhile, engineers relied on teleoperation, which was particularly useful to space exploration.

The U.S. and Soviet space agencies had both committed to robots and lunar vehicles early in the space race. By the early 1970s, two Soviet LUNOKHOD wheeled rover/laboratories had successfully collected samples, completed experiments, and sent back thousands of images to Earth. As part of the U.S. Apollo Moon mission research, the Surveyor program sent probes with robotic laboratories to the moon in the mid-1960s, in part to see if astronauts and other vehicles could move around on the surface. The STANFORD CART, a simple platform driven by four bicycle wheels and a bike chain, was originally built during the Apollo Program to see whether an unmanned vehicle on the moon could be teleoperated via radio signals from Earth. This goal was temporarily shelved when NASA modified its mission plan and contracted Boeing to produce lunar roving vehicles (LRV) to be controlled by astronauts. With the CART idle, at least three students "adopted" it for their graduate research.

The STANFORD CART

In the late 1960s, John McCarthy, founder and director of SAIL, discussed with his student Rodney Schmidt the idea of replacing the emotional and sometimes irrational human automobile driver with a more efficient, safer computer-controlled navigation system. Schmidt took on as his thesis project the problem of building a vehicle that could navigate autonomously in traffic along unmodified paths using onboard sensor information.

In this phase, the CART diverged from the engineering philosophy of the SHAKEY project by getting its initial understanding of the world directly from its sensors (rather than from a preprogrammed IWM) and using those readings to modify its physical behavior. At the end of Schmidt's research project, the CART could follow a white line down a road for a short distance. It had not only broken away from the preprogrammed IWM, it had done it outdoors in an unprotected environment. Bruce Baumgart, the student who took over the CART in the early 1970s, argued that if mobile robots were going to be any good outside the protection of flat-surfaced labs, they would need more comprehensive perceptual skills. He spent his graduate research tenure developing an accurate 3-D graphical interface to model the CART's environment. Although Baumgart's 3-D research was innovative, it took so much time that little of it was ever applied to the CART. (Brooks, 2002, 27)

Hans Moravec, later head of the AI Lab at Carnegie Mellon University (CMU), took on the CART as a project while he was a student at Stanford in the 1970s. Arguing that in nature there is a direct link between imaging eyes, large brains, and mobility, he rejected a preprogrammed symbolic world model. Like the earlier versions, Moravec's version of the CART planned its own path based on the visual information it collected. His approach was to mount a multipositional camera that he would adjust several times to let the robot gather visual information from different angles. It would then combine this information to get its 3-D model, process the information, and then move again. If it perceived new obstacles along the way, it could modify its plan, thus navigating through cluttered spaces. Unfortunately, because it had to stop and process the visual information it acquired, it lurched ahead only about 1 meter every 10 to 15 minutes. Despite failing some of its trials, Moravec's mobile robot did successfully traverse several 20-meter courses. Beginning in 1981, Moravec continued to work on this problem with the CMU Rover, a project funded by Office of Naval Research (ONR).

Other mobile robot research of the period took a similar approach. For instance, a three-wheeled robot was developed at the Laboratoire d'Analyse et d'Architecture des Systèmes (LAAS) in Toulouse, France, in the late 1970s. Using sonar and laser range finders to help build two-dimensional models of its test world, their robot, HILARE, successfully navigated between moving plywood walls. Like other mobile robots of the period, it would collect the sensor data, build a plan of how to move while avoiding obstacles, and then implement the plan.

GROWTH OF THE FIELD

In an early 1980s' assessment of the state of advanced automation for space missions, NASA articulated its concern that regardless of the aggressiveness of AI and robotics research, the portion devoted to space missions was inadequate to the needs of the agency's goals for the coming decades. Furthermore, it perceived a scarcity of qualified workers in the AI field. These same concerns were voiced by other organizations involved in advanced automation, and consequently, representatives from government, business, and engineering began a series of programs to encourage growth of the related fields, including competitions and direct research funding and recruitment. By the middle of the 1980s, agencies like DARPA, NASA, ONR, and corresponding international agencies had stepped up support for research on new robotic platforms that could be adapted to warfare as well as space exploration, oceanographic research, and health care. By the end of the decade, CMU had founded the world's first doctoral program in robotics, and dozens of graduate students in computer science departments at major institutions were working on robots for their thesis projects. Improvements in the complexity and speed of microchips, and the introduction of neural net programming and massively parallel processing (MPP) supported the expansion of robotics over the next two decades. This was demonstrated in sophisticated sensor technology used in new iterations of the mobile robot, just a few of which are discussed here.

A significant portion of research focused on the problem of real-time response. Although using information directly from sensors for "on the fly" path planning is an improvement over preprogrammed IWMs, the model can still become outdated while the computer is processing information from the robot's sensors. For instance, someone may unexpectedly turn a corner toward the robot, liquid could begin dripping from the ceiling, or another robot could move into the path. While the robot stops to process this information and modify its plan, the situation can change again. Another problem is that sensors still were essentially only reading edges and judging range. Even if the obstacle was something as benign as an empty cardboard box, the robot might not know it could push it aside to make room to pass between the box and the wall. Stanford Research International (SRI) built FLAKEY to solve these sorts of problems.

FLAKEY

In 1984, SRI, by this time divested from Stanford University, began testing a second-generation mobile robot with real-time stereo vision algorithms to distinguish and follow people, and the DECIPHER speech recognition system to respond to spoken commands. FLAKEY, programmed by Rosenschein and Kaelbling, responded to the world using an SRI innovation that combined sensor feedback with map-based operating information. Their

Local Perceptual Space (LPS) module was a repository of information like raw distance readings, extracted surfaces (e.g., obstacles), objects from the map (e.g., specific corridors), and abstract control points from the strategic plan (e.g., positions to obtain), posted along with interpretation routines.

Programmers made it possible for FLAKEY to perform both purposeful behaviors such as following a corridor to reach a target location, and reactive behaviors like avoiding obstacles by providing it with a set of fuzzy rules (see Chapter 4). When the precondition of a rule is present, its action is interpreted as highly desirable; whereas if a precondition matches only partially, the action is interpreted as less desirable. Blending the actions of all the rules according to their levels of activation results in an "optimal" action. PRS-LITE provided real-time supervisory control for FLAKEY, sequencing the robot's goals using a library of predefined procedures. High-level goals were reduced into subsequences of lower-level goals where behaviors are activated and deactivated according to the refined plan and the situation-dependent information in the LPS.

It took almost a decade to achieve this level of coordinated sensor-processing technology, but the team that worked on FLAKEY won awards for these innovations in both the 1992 and 1993 AAAI mobile robot competitions. Additional visual recognition of landmarks and visual stereo processing for obstacles had been incorporated into the robot for the 1993 contest, where it finished in the top category in the two events it entered. While FLAKEY was in development, researchers made progress in autonomous robots that could maneuver over rugged terrains.

Autonomous Land Vehicles

Another area where real-time response is crucial is in unstructured environments. ALVs must be able to maneuver in inhospitable weather and rough terrain. Consequently they sometimes move on tracks rather than wheels alone, in order to help the robot surmount hilly or rocky environments. Keeping in mind the military's mission to send unmanned supply and artillery vehicles into war zones, engineers worked to develop systems that could sense their surroundings, navigate to a specific goal, and execute their mission without human intervention. This involved the development of integrated systems of advanced sensors, data analyzers, power controllers, and backup power systems. During the 1980s, a number of government and commercial entities around the world were pursuing similar goals. For instance, Bundeswehr University developed an ALV project funded by German automobile and electronics companies. A major U.S. contributor to this initiative was CMU NavLab, launched in the mid-1980s under the direction of Chuck Thorpe, one of the first CMU alumni to pursue a career in robotics and now director of the CMU Robotics Institute.

In 1984, as part of its Strategic Computing Initiative, DARPA funded ALV research to develop robotic vehicles for civilian and military use. While pursuing his doctoral degree in the Computer Science department at CMU, Chuck Thorpe wrote a proposal to DARPA describing techniques for driving under hazardous conditions. As a result, the agency funded the first NavLab, a Chevrolet van equipped with a $1/2 million of computer equipment. Despite the fact that the van could only travel at about 1.5 miles per hour, the lab received additional funding, producing another 10 prototypes over the next two decades. Cars, buses, HUM-Vs, and trucks were modified to test teleoperation, vehicle control on rough terrain, and a variety of sensing techniques including the use of color classification and neural nets in road-following systems. For instance, ALVINN was a back propagation neural network system that learned to navigate autonomously by watching people drive. It viewed the scene through a two-dimensional, 30×32 unit artificial retina that received input from the vehicle's onboard video camera. Each input unit was connected to a layer of five hidden units, which are connected to a 30 unit output layer, a linear representation of the best path to follow.

The researchers also tested a variety of three-dimensional terrain representation approaches including obstacle maps, terrain feature maps, and high-resolution maps. Their work in driver-performance modeling and traffic simulation earned them an invitation to take part in the U.S. Department of Transportation (DOT) 1997 National Automated Highway Systems Consortium demonstration in San Diego, a showcase to demonstrate the technical feasibility of automated vehicles. The NavLab buses and cars drove themselves in the HOV lane of Interstate 5 at speeds between 60–100 miles per hour.

Environmental Cleanup

An important byproduct of government-sponsored, university-based research, are the many spin-off companies that were established to develop and commercialize lab innovations. One of the first companies to produce robotic equipment for dangerous environments was RedZone Robotics Inc. founded in 1986 by William Red Whittaker, director of the CMU Field Robotics Lab and a principle investigator on NavLab I and II. Whittaker had been one of the first engineers to pursue the use of robots for inspecting and cleaning up biohazards after the nuclear disasters of the late 1970s and 1980s. In 1984, robots developed at RedZone were used in the cleanup of the Three Mile Island disaster (1979). RedZone's robots also assisted in the cleanup of the Nine Mile Point Nuclear Station, and in the remediation and closure of underground nuclear waste storage tanks at Oak Ridge National Laboratory. They were eventually contracted by the Department of Energy (DOE) to produce a mobile teleoperated robot to analyze and repair the decaying sarcophagus built around the Chernobyl Unit Four nuclear reactor that had exploded on April 28, 1986. Their radiation-tolerant PIONEER was designed to be operated remotely, negotiating narrow passageways and collecting images to build a 3-D model of the conditions. Virtual reality software originally developed at NASA's Ames Research Center for the Mars Pathfinder missions was donated to generate the 3-D models of the information sent back by PIONEER. Silicon Graphics donated almost a half-million dollars in computer equipment, including an ONYX-2 server and three Octane workstations. PIONEER was to be tethered to one unit to generate and analyze the maps, while another was stationed at the Chernobyl administrative building. The third Octane was set up at the University of Iowa to monitor the task via NASA satellites. After successful test runs, there was no further action by PIONEER for some time. Some have speculated that political and economic upheavals in the Ukraine, along with differences of opinion about ownership of the robot kept the project from continuing. However, after 2001 some reports were issued from the Ukraine noting that since ownership had been transferred to their country, operators had been learning to use the equipment and it was expected that PIONEER would be used for inspection and cleanup of a number of compromised sites.

HELPMATE

The financial potential for a service robot industry was not lost on industrial robot manufacturers, who had the advantages of financial capital and sophisticated tool shops in which to develop them. In 1984 after selling Unimation to Westinghouse, Joseph Engelberger founded HelpMate Robotics Inc. in Danbury, Connecticut, where he served as the company's chairperson until it merged with the Pyxis Corporation of San Diego and Engelberger retired. In 1999 HelpMate was acquired by Cardinal Health. During that time HelpMate engineers produced innovative vision and navigation technologies including an improved light direction and range (LIDAR) scanner. LIDAR is a device in the "eyes" of the robot that senses light, calculates direction, and determines the range to objects in its path—an improvement over sonar. Researchers also improved self-navigation capabilities by combining data from a variety of innovative sensor technologies. They achieved reliable robot control in areas with predefined, fixed components such as doorways, light fixtures, windows, and elevators that are definable from photos or engineering drawings, as well as near unexpected objects, such as a patient on a gurney or people moving around the space.

The HelpMate® trackless mobile platform was well suited to the health-care industry, where it was configured to a variety of relevant tasks, like transferring medications or delivering meals to patients. Since the late 1980s a number of mobile robots have been developed to meet the needs of the hospital system. For example, the RW II was developed at CMU in conjunction with the Magee Women's Hospital in Pennsylvania. RW II was a stand alone, autonomous robotic workstation meant to replace high-paid technical workers in tasks such as decontaminating surgical instruments, moving specimens in laboratories, transporting medications from the pharmacy to the nursing units. Since that time mobile health-care robots have expanded into a major market as a cost-effective way of relieving hospital staff of tedious tasks that keep them from focusing on direct patient care. A number of these will be discussed in Chapter 6.

FROM THE BOTTOM UP—BEHAVIOR-BASED ROBOTICS

Advances in sensor and computer processing technologies were advantageous in many settings like those described above; but there was a gap in the availability of robots geared to situations where either budget or workspace was limited. By the mid-1980s, some researchers were considering alternative approaches to fulfilling the now decades-old promise of building human-level AI. With the availability of dense microchips a number of scientists turned to neural nets, an artificial analogy to the structure of the human brain useful for machine learning (see Chapter 4). Artificial neural nets helped computers and robots learn their jobs quicker. Other researchers, drawing on contemporary work in evolutionary biology and the behavioral sciences concluded that intelligence is more than winning at chess or doing differential equations, or even processing and internalizing complex images and other information. It is the ability to adapt and survive in a continually changing environment. They directed their research away from building computers that *think* intelligently toward building useful machines that *act* intelligently. This biologically based approach would come to be known as Behavior-Based Robotics (BBR).

BBR begins from the assumption that biological organisms possess attributes that can be explained in mechanical terms, and that conversely, traits like self-preservation and intentionality can be implemented in a robot as a senor–actuator control system, just as that for lifting and placing an object. The basis of this assumption can be traced back to the man–machine philosophy of the Enlightenment, and more recently to the work of Wiener, Walter, and Ashby. One of the first proponents of BBR was Rodney Brooks, today director of MIT's CSAIL (a merger of the AI and the Computer Science Labs) and cofounder of the Massachusetts-based iRobot company. Brooks was already a veteran of the university research system when he became a faculty member at MIT in 1984. He had assisted Hans Moravec with his CART project while they were both graduate students at Stanford. As a postdoctoral scientist at CMU and MIT and junior faculty member at Stanford, he had spent a decade following the existing paradigms, working on computer models of the world for industrial robots, computer vision, and path planning for mobile robots.

Once he was on the faculty at MIT, Brooks began his research into BBR. His first experiment ALLEN (named for Allan Newell, a cofounder of AI) was a 25-centimeter-high cylindrical robot mounted on a three-wheel mobile platform developed by a young engineer named Grinnell More. Its primary sensing device was a 12-sonar ring developed by Polaroid as an auto focus mechanism, which Brooks mounted atop the cylinder along with two cameras to give ALLEN a comprehensive view of its environment. A serial line communicated the robot's directional and velocity commands. Brooks had to link ALLEN's single on-board processor to the mainframe, a LISP machine, with a 20-meter-long cable.

Frustrated with the time-consuming task of programming computational control systems, he analyzed the current paradigm in basic terms:

Sensor →	Computer →	Plan →	→ Action
data	builds model	sent back to robot	
	and plan		

In the current paradigm, it seemed to him that most of the robot's time was spent in the computation stage. Brooks returned to the biological analogy. Animals don't have to stop and think between each step or turn of the head; why should robots? He proposed getting rid of the computational "middle man", leaving:

$$\text{Sense} \rightarrow \quad \text{Act}$$

According to Brooks, if perception and action are the keys to more general intelligence a robot needs to be both situated and embodied. Unlike computers, embodied robots have the physical ability to interact with their environment. Situated robots get their information about the world not from abstractions but through direct sensorial experience. The problem was how to engineer such a robot.

In nature, organisms develop over long periods of time and at various stages, and abilities that are more advanced grow over the old ones. In other words, nature builds intelligence into biological organisms in layers. For example, the human brain in all its complexity evolved "over" the primitive reptile-level brain, which still functions in us for low-level reasoning. Brooks theorized that mundane activities like walking, searching, or avoiding obstacles are the foundation of higher level cognition used for problem-solving mathematics, and philosophy.

He built ALLEN according to this reasoning, producing a far simpler form of programming than the highly symbolic system Rosenschein and Kaelbling had used for FLAKEY. He called this "bottom up" engineering approach to BBR *subsumption architecture.*

First, simple control systems were built for ALLEN's most basic behaviors; then more sophisticated control systems were added for more complex behaviors. In this system, if the more complex layers "know better" how to do something, they will "subsume" the lower level control. For example, ALLEN's lowest level program was obstacle avoidance. Layer two was the motivation to wander aimlessly. There was no need to worry that the robot would bump into things in its travels, because the lower level avoidance programming was in place already. Its third layer directed ALLEN to move in the direction of something interesting (like Grey Walter's tortoises looking for light). Brooks expressed ALLEN's nervous system as a circuit diagram that connected a dozen or so simple computational elements with sensor input and actuator output. ALLEN was eventually equipped with more layers that allowed for more methodical exploration.

Brooks unveiled his new paradigm at the 1985 International Symposium of Robotics Research in France using videos of ALLEN traversing the halls briskly, without pausing. ALLEN did not look much different from other early mobile robot experiments, however the robot operated with one small but significant difference: there was no computation module. Brooks had removed the step of relaying sensor data to the computer to build a model and plan and relaying it back to the actuators. He had demonstrated that a robot could act with purpose and intelligence by communicating sensor activity directly to actuators. Furthermore, he had demonstrated real-time movement using layered reasoning rather than mapping. Whereas a computational-based AI encountering an unexpected situation will hesitate and may not be able to complete its task, a BBR will keep exploring because that is what it is built to do. It has no preconceived notion of its world, only a basic motive to go out and look for things. "Mapping" is not in its vocabulary.

Although the published version of Brooks's presentation is now often described as groundbreaking and many robots are built according to this paradigm, at the time it received negative peer review. Engineers were hesitant to accept the notion that the big, complex, time-consuming work they were doing to build intelligence into their machines might be unnecessary. Others thought it could be useful in very specific situations, but not for really complex tasks. Confident that he had solved a major engineering problem, Brooks followed ALLEN with HERBERT, named in honor of Allen Newell's coauthor of LOGIC THEORIST, Herbert Simon. All of HERBERT's sensing equipment, including 8-bit processors, an arm, and laser scanner to collect environmental information, were onboard. HERBERT was situated in the real world. It got its information not from an abstraction, but from experience. The robot's "job" was to locate and dispose of empty soda cans, which it did successfully, discerning between empties and those that still had liquid in them. Although the activity was simple, it was the first example of a BBR performing a service autonomously. HERBERT, and other student-built BBRs in that lab used very little internal memory, yet they were able to successfully maneuver around the building, and do simple tasks, all using very low-level search-find-execute programming.

Brooks argued that by removing the pricey and time-consuming computational element BBRs could be built quickly and economically. The idea was summarized with the axiom, "Fast, Cheap, and Out of Control," the title of a paper he published with Anita Flynn in 1988 to provide a rationale for sending BBRs on space missions. They reasoned that sending a number of smaller rovers to explore would cost far less in production time, and in programming, lifting, and deploying one truck-sized robotic vehicle.

Although the initial reaction to the idea of using such simple robots on other planets was not particularly enthusiastic, the potential of BBR was eventually realized in the space program. According to Brooks the concept made its way into NASA by a circuitous route. During a summer research internship at Jet Propulsion Laboratory (JPL) in 1989, MIT student Colin Angle (later a cofounder of iRobot Company with Brooks) produced a small, four-wheeled robot with a gripper made from Radio Shack parts using subsumption architecture. At the time, the best prospect JPL had for putting a rover on Mars was a one-ton prototype called ROBBIE, and an estimated budget of $12 billion. Naturally Angle's "fast, cheap, out of control" rover, TOOTH attracted the attention of the Mars mission manager, Donna Shirley, who provided a small budget for further research.

This was used to build the ROCKY series of prototypes that resulted in the SOJOURNER Mars rovers described in Chapter 6.

Wheeled and track robots have been very successful to the expansion of robotics. Three- and four-wheeled vehicles are generally quite stable. As long as their center of gravity is not too high, they will not tip over in normal movement. However, a number of researchers thought that rather than continually reconfiguring vehicles to move into human environments, it would be advantageous to build robots that could walk into those spaces. Legged animals have the ability to step over and around obstacles on a smaller footprint than wheels or tracks. Still, this presented researchers with the challenge of balance and stability.

The Leg Lab

Marc Raibert is one of the pioneers of legged robotics in the West. He founded the Leg Lab at CMU in 1980 to study the principles of animal locomotion, then moved it to MIT in 1987, where it was established as a subgroup of the AI Lab. The initial project, funded by Ivan Sutherland and then ARPA, was a pneumatically powered, planar, one-legged hopper used to investigate dynamic balance and stability. A computer controller monitored its actual speed and position, and adjusted the angle of the leg to keep the robot upright. The leg adjusted its balance with each hop. It could hop in place, travel at specified rates of speed, and maintain its balance if disturbed. Over the next two decades the lab tested a series of planar and 3-D one-, two-, and four-legged units using generalizations of the original hopper. The quadruped (1984–1987) was able to move by trotting, pacing, and bounding. These particular gaits were chosen because they involve pairs of legs. The idea was to transform the focus of control to a virtual biped gait. By 1986, the planar biped could do flips, aerials, and run at 11.5 mph, and two yeas later it was using selected footholds to climb a short stairway. In 1993 the team developed UNIROO, a mechanical analogy to a kangaroo that consisted of a simplified body, a three-joint articulated leg (hip, knee, ankle), and a single DOF tail. This experiment demonstrated that it is possible to control the balance of legged robots that have a nonsymmetrical mechanical structure. (It was Raibert's lab that supplied the UNIROO and 3-D biped seen in the 1993 film, *Rising Sun*.) In 1997, a planar bipedal robot, Spring Flamingo, demonstrated walking using actuated ankles and feet using sensors in its ankles to help it balance.

Quadrupeds

The Leg Lab was not the first lab to produce a quadruped robot. Ralph Mosher had developed a four-legged walking truck for General Electric in the 1960s that was meant to load bombs into bomb bays. In tests, it could load up to 500 pounds moving at 5 mph. It had articulated legs; however it was essentially an exoskeleton. A human operator inside the vehicle had to maneuver the legs via joysticks like a crane operator. This proved extremely fatiguing for the operator. Furthermore, the fact that the operator was required to adapt to the machine contradicted the point of robotic assistance and the project was abandoned.

In 1981 while Raibert was working on the one-legged hopper, Shigeo Hirose at the Tokyo Institute of Technology produced the TITAN II quadruped. It was the second in a series of TITAN robots developed in the lab over the next 20 years. TITAN II could feel its way up stairs of varied heights using contact sensors located on the sides and bottom of its feet. It was engineered to translate small motor movements inside its body into larger more complex leg movement. The system allowed the quadruped to respond to touch with animal-like reflexes. Four-legged robots were lighter and more energy efficient, but were unstable compared to six-legged ones.

AMBLER

Red Whittaker built a six-legged prototype for exploring the surface of Mars during the 1980s, however he set aside the animal analogy used at the Leg Lab and retained the computational module that Brooks had rejected. His Autonomous Orthogonal Legged Walking Robot (AMBLER) was designed to operate semiautonomously. Remote human operators provided the robot its target location, while the robot itself planned the trajectory and step sequence. Anticipating that the robot would encounter obstacles like large boulders, deep crevices, and steep slopes, he designed AMBLER to be 3.5 meters high, with the ability to step over objects up to 1 meter high. To decrease the amount of sensing and planning (computation) necessary for articulated legs in such a large robot, his team built AMBLER with unbendable, telescoping legs.

The legs remained vertical while they swung horizontally, then telescoped vertically to touch the ground. Since they did not rock or sway in the act of stepping, unnecessary collision with obstacles was avoided. Unlike multilegged biological organisms, AMBLER could step with any leg in any sequence, and move its rear-most leg past all others, covering extreme terrain more efficiently. AMBLER remained upright through even the roughest terrain by keeping its legs vertical and its body horizontal. The stability achieved through this straight-legged configuration as well as the high vantage point of its range finder allowed the robot's perception system to build smoother, more comprehensive terrain maps. The robot's complex movement was controlled through neural networks to learn the optimal footholds for particular terrains, a gait planner to evaluate both the condition of the terrain, limits of movement, and to coordinate body and leg movement, and Task Control Architecture (TCA) to coordinate all the systems.

The problem with vehicle-sized robots with six or eight legs is that they are heavy and require immense amounts of power for each leg to move. Whittaker used his patented walking method to reduce power consumption. However, while a 20-foot-tall machine could step over crevices, it could not work *within* them without the addition of manipulators with specialized endeffectors that renew the weight and energy problem.

GENGHIS

In the 1980s Brooks and his students were working on a series of small multilegged walking robots. Generally, the more leg pairs an insect has, the better it can stabilize itself under changing terrain conditions. Insects, which use *static* balance, have become a popular biological analogy for building small, lightweight, nimble walking robots. Under normal circumstances, a six-legged creature can move easily using a *tripod gait*. In this case insects move the front and rear legs on one side and the center leg on the other side at the same time, thus keeping a "tripod" of support at all times. When the creature lifts a leg from one side, another from the other side is set down. Brooks reasoned that insects operate in the world with only a fraction of the number of neurons present in human brains; yet they are able to move around several feet in just seconds while avoiding obstacles and reach their goal. They even stumble from time to time, and manage to regain their balance and move on. He thought this should be possible with BBRs as well. Rather than building them to be always stable, subsumption architecture would allow them to do what insects do. He built GENGHIS on this principle.

Unlike AMBLER, which had to plan each of its leg movements, GENGHIS—like ALLEN and HERBERT before it—explored with little more plan than to look for something interesting. In the case of GENGHIS, the robot looked for heat. It had a six pyroelectric sensor array mounted up front, which allowed it to sense body heat. Therefore it would follow people around. GENGHIS was entirely situated with all its mechanics onboard. Its subsumption-type programming was done with 51 *augmented finite state machines* (AFSMs) running in parallel; these mainly controlled the steer instruction, and action of its legs, which moved either up and down or back and forth from the "shoulder" joint, allowing it to scramble over or around anything in its way. It didn't need programming to follow people, since the sensors were pointed forward, it moved forward in the direction of the heat automatically.

DANTE

In the early 1990s NASA sponsored a competition for the development of lightweight, nimble, mobile robots to explore inhospitable environments. The long-range goal was extraterrestrial exploration, although research was also applicable to hazardous terrestrial activities. Antarctic volcanoes were chosen as test sites for their similarities to the Martian environment. The CMU Field Robotics Center (FRC) developed an eight-legged robot DANTE I that explored the inside of the Mount Erebus volcano. The frame-type walking robot carried a lightweight, compact, durable multiprocessor control system that was responsible for coordinated motion control, foot-fall sequencing, science control, dead reckoning, image collection, temperature regulation, and system safety. From December 1992 through January 1993, DANTE rappelled into the active volcano to retrieve pristine gas samples from near the convecting magma lake. The importance of robots for exploration of hazardous terrestrial and extraterrestrial environments was demonstrated in 1993, when eight volcanologists were killed during two monitoring/sampling missions. Continued research resulted in CMU's improved six-legged tethered walker, DANTE II. In July of the following year, it rappelled down the crater walls of the Mt. Spurr (Aleutian Range, Alaska) volcano taking fumarolic gas samples sending back images over Mosaic, a predecessor to the World Wide Web. During the last several years, research in mobile robots has resulted in a number of new configurations that will be discussed in Chapter 6. However, some of the most interesting research was begun during the latter part of the twentieth century. It resulted in the first humanoid robots.

Humanoids

Much of the work done at the Leg Lab was aimed at understanding and simulating bipedal locomotion in machines. Bipedal movement is more difficult to translate into machine motion than multilegged configurations because it requires *dynamic* balance. That is, one foot is always off the ground during walking, and during running or jumping gaits there is a point where both feet are off the ground. Once healthy human children are a few years old, they achieve autonomous dynamic balance and can climb stairs, step over small objects, run and jump intuitively. Humans developed this capability over many hundreds of thousands of years. It was therefore a considerable accomplishment that robotics engineers achieved this capability in robots in about 30 years.

Dynamic balance in bipedal robots requires coordinating the mass and weight, and placement of the robot body with its legs. The torso, head, appendages, and actuators as well as any batteries, equipment, and processors must be considered. This *payload* must be distributed so that the robot's center of gravity is not too high (or it will tip over), and so that the weight does not exceed the power and strength of the grounded leg. A solution to the problem of bipedal balance and control, *Zero Moment Point* (ZMP), was proposed in 1970 by Serbian mechanical engineer Miomir Vukobratović. This is the point where the combined forces of gravity and inertia working on the robot intersect with the ground. ZMP was the basis of the first walking humanoid robots, which were produced in Japan in the early 1970s.

Japan has a very low birthrate and no significant immigrant population that in other nations would take on the low-pay, tedious work. In anticipation of a rising geriatric population and limited labor force, Japan committed to robotics just after World War II, and began working on humanoid robotics in the late 1960s. Their philosophy is that since service robots will work closely with human beings within the normal human environment, the closer they are in appearance and behavior to humans, the better humans will respond to them. In addition, the human form is more suited to many situations than other machine configurations.

The Waseda Humanoids

Since the 1980s, dozens of labs in Japan have been working on different aspects of the humanoid concept, however the first humanoids were produced at Japan's renowned Waseda University. The bioengineering research group of the School of Science and Engineering at Waseda (now known as the Humanoid Research Group) produced Waseda Robot-1 (WABOT-1) in 1973 under the direction of Ichiro Kato. Around 1984, they produced WABOT-2, which performed music daily at the Japanese government Pavilion during the Science Exposition in Tsukuba in 1985. Although it was bolted to its seat it did have legs and feet, which manipulated the pedals of the instrument. WABOT-2's computational and sensor system worked together. According to the reports, it was not just manipulating the keys through a program of movement, but reading the music, processing what it saw, and responding by playing the notes.

Thereafter Japan expanded its work in this area and began a humanoid research institute. Waseda became involved with the Humanoid Project, a consortium of Japanese government, academia, and industry whose objective is to develop a humanoid robot that "shares information and behavioral space with humans" (Hashimoto, 26). The research is divided into four areas: vision, speech, brain, and locomotion. Together, the project produced HADALY-1 (1995), a prototype that demonstrates human-robot interaction. More recently, the Humanoid Project was admitted to the New Energy Development Organization (NEDO), within which HADALY 2 and WABIAN were developed. WABIAN is a complete bipedal humanoid that can carry objects like a human. It is described as capable of functioning in the home in a communicative way.

HADALY-2 is a social robot. The project focus is on conversation (voice recognition and synthesis), vision, and gestures. The engineers claim it possesses physical gesture and conversational capabilities that make it safe and gentle for interacting with humans in a home environment. The purpose of these two projects is to provide all the robotic ancillaries necessary for robots to live among humans. These technologies were developed in the seven laboratories associated with the Humanoid Project. The engineers reiterate the purpose for developing these robots as the close emotional and physical support of humans in eldercare, convalescence, assistance around the home, and other service areas. (Hashimoto, 26) Early iterations of the humanoids in Japan were tethered to their processors and power sources, just like in the West, though in recent years, some of them have become more autonomous.

The commitment to the humanoid paradigm in Japan has to a certain extent been an homage to the late Kato, who disseminated his philosophy among his students during three decades. They in turn carried the philosophy into other university labs and into commercial robot companies. Kato espoused the ideas of mechatronics, the replication of animal function in mechanical systems, and *synalysis* (from synthesis and analysis), the idea that the replication of human function in machines requires that we first analyze and understand human function and behavior. This philosophy is now present in many robotics labs in the West, where proposals for humanoid projects are often described as an investigation into human neurology, kinetics, or behavior. It is uncertain whether the concept emerged in both hemispheres independently, or whether it migrated from Japan through the media and through the professional movement of graduates of the Japanese university labs. In either case, by the late 1980s, humanoid projects began to appear in the West.

One of the first was the United Kingdom-based Shadow Robot Company, whose mission statement posted to the World Wide Web in the 1990s asserted that it was developing the humanoid to relieve human beings of household tasks so that they could pursue more enlightening activities. The engineering group, which has been developing robots and robot components since the late 1980s, argues along with most proponents of humanoid technology, that building machines that can navigate human environments like stairs and kitchens in the same way that humans do would be far more efficient than reconfiguring human environments to accommodate bulky service robots. Furthermore, they offered the popular rationale that the humanoid configuration is one which has already been tried and tested by Nature, and has proved successful. Shadow has recently suspended its SHADOW BIPED project, but this may be due to the company's success developing and selling associated robot parts, like the artificial muscles and hands discussed in Chapter 6.

COG

Among the most famous humanoid robots is COG, a shortened form of "cognitive" and a reference to wheel cogs, which Brooks describes as a merging of the mechanical and intellectual. This BBR, begun by Brooks around 1993 and continued over the following decade by Brooks and his graduate students, was developed to learn by trial and error like an infant, rather than by following complex programs. It was based on two assumptions. First, human beings' body configuration contributes to the experience from which we develop internal thought and language. Thus for a robot to gain human-like intelligence, it would have to have a human body so that it could develop the same type of representations of the world. Second, part of what makes us human is our interaction with others. A humanoid robot would make it easy for humans to interact with it, and for the robot to learn social cues.

In order to make COG behave like a human and to contribute to the new idea of sociable robots, it would need to develop human social skills related to vision as recognition, attention, choice, and turn-taking. It would also need dexterity and coordination of its limbs. COG has been controversial because the project emerged at the same time as Doug Lenat was developing CYC® (discussed in Chapter 4) and a debate ensued over whether it was better to instill intellect in robots using global databases or by using the behavior-based approach to learning. The scientists who have worked on COG over the past decade or so have used theories of human cognition, early childhood training, linguistics, and sociology to instill in COG a sociable nature, and this work parallels the inroads made in social robots in Japan. Robot learning became a central area of investigation over the last decade, and the variety of approaches and innovations will be discussed in the following chapter.

Staying Alive

By the end of the twentieth century, robotics had become a multibillion dollar industry, providing robots to businesses, government and private service agencies, and consumers. A new generation of researchers absorbed the theories of their predecessors in AI and robot engineering and produced a wide variety of autonomous and semiautonomous intelligent machines that rival the robots of science fiction. Still, a number of critical issues remain, among them power. Industrial robots can easily be continually connected to electric or hydraulic power systems. Mobile robots cannot be autonomous unless they have a continuous onboard power supply. Researchers today count the lack of efficient power sources as a central challenge to the production of autonomous robots. The Honda humanoid, ASIMO, runs on a 40-volt nickel metal hydride (NiMH) battery that provides only around an hour of operative life for its complex system. The recharge time for full battery power is 4 hours, although Honda has been working to improve that record. Likewise, Kawada Industries' HRP-2 android's NiMH 48-volt battery only lasts about an hour doing complex tasks. Other humanoids, like the Tmsuk service robot are presently teleoperated via computer or cell phone.

One alternative for mobile robots has been to set up protocols that allow robots to plug into a recharging unit autonomously. This was the protocol first used by the HOPKINS BEAST in the 1960s. It was more recently used by SWEET LIPS, one of a series of mobile robot tour guides produced at CMU in the 1990s and field tested at the Carnegie Museum of Natural History in Pittsburgh, Pennsylvania. The robot guided museum visitors and instructed them about the exhibits through an interactive video setup in its "chest." At the end of each day, it had to maneuver back into a power den and plug in to recharge. Researcher Milo Silverman at USC points to the limitations of this method. For instance, in the case of the mobile robot, CYE, which comes with its own recharging station, if the robot cannot locate the station, it shuts itself down and is useless without human intervention. In the early 2000s, Silverman's team produced a more robust system with a user-monitoring interface, making their robot capable of an autonomous docking/recharging.

The decision to develop humanoid robots as personal assistants, or "partners" as some Japanese companies now call them, requires attention to the issue of dexterity. We take for granted how important the tactile sense is in our lives. Without even seeing an object, our sense of touch allows us to discern whether we are handling a piece of paper, a rock, or a knife. These are distinctions of material: solidity, density, sharpness, weight, and even moisture-level and temperature. Consider what kind of financial loss a manufacturing or packaging business would suffer if a worker—human or robot—could not tell if the object to be picked and placed was a paper cup and not a golf ball or a steel wrench. In only moments, thousands of pieces of product would be crushed or dropped and destroyed. Therefore, accurate grasping, handling, holding, and passing objects of different sizes, weights, and densities has been a central concern for industrial robot engineers from the beginning of automation. It has also been a key concern for service robots meant to operate in remote, rough locales like disaster sites, undersea wreckage, and on other planets. This becomes even more crucial in situations where a robot is working closely with a human, perhaps handing objects back and forth, or lifting patients. A number of labs have produced highly dexterous, precision hands. Among them are the Shadow Robot Company and JPL, whose teleoperated ROBONAUT is meant to work in space doing extra-vehicular activities (EVA) like making repairs to the International Space Station (ISS).

Improvements in artificial muscles, vision systems and audio, speech recognition and synthesis, haptic sensors, and synthetic skin have contributed to all areas of robotics, and in particular to the advance of humanoid robotics. In the past 5 years, innovations that extend robots' ability to perceive and manipulate their environment have made it possible for them to operate autonomously in more situations. In the final chapter, I describe a number of the most-recent innovations related to robotics. I will also show how robots are entering our lives and gaining acceptance as household helpers and museum guides, and how media coverage is creating anticipation for them. Finally I will describe how a new generation of robot engineers is being recruited through toys, games, and competitions.

6

Robots among Us: The Latest Developments

◆

From a largely dominant industrial focus, robotics is rapidly expanding into the challenges of unstructured environments. Interacting with, assisting, serving, and exploring with humans, the emerging robots will increasingly touch people and their lives.

—Brian Carlisle and Oussama Khatib
2000 IEEE International Conference On Robotics and Automation

At the beginning of the third millennium robotics was expanding despite periodic declines in sales during the previous decade. The annual joint survey by the International Federation of Robotics (IFR) and The UN Economic Commission for Europe (UNECE) reported an estimated 742,500 industrial robots in use worldwide in 2000. In the first half of 2005, orders for industrial robots rose by 13 percent above the same period in 2004, with sales to North America (United States, Mexico, Canada) showing an increase of 36 percent for the same period. Meanwhile, markets for professional and personal use (domestic) service robots had emerged. According to IFR/UNECE, by the end of 2003 there were 607,000 domestic robots in use worldwide, with the majority of sales in robotic lawn mowers (570,000 units) and the increasingly popular floor vacuums (37,000 units). For instance, in October 2004, the Massachusetts-based iRobot Company reported that sales of its low-cost ROOMBA vacuum had reached one million units in two years. The joint survey estimated that by the end of 2004, at least 25,000 professional-use (underwater, medical, forestry, farming, defense, public relations) service robots had been installed. They predict that by 2008, the service robot market will meet or exceed that of industrial robots and that 4.5 million of those will be domestic use robots.

At the moment, industrial robotics still dominates the field, with automotive manufacturing and parts production representing the largest share of the market. According to the American organization, Robotics Industries Association (RIA), orders from automotive manufacturers increased by 49 percent during 2005, while automotive components companies increased their orders for robots by 14 percent. These two sectors accounted for 70 percent of new robot orders in 2005. Statistics vary from organization to organization based on differences in classification and method of reporting, but it is generally agreed that 2005 was a record year for robotics.

Sales of robots to automotive *parts* manufacturers increased again in the first quarter of 2006, although there was a 30 percent drop in orders from automotive manufacturers in North America. This decline was expected in light of a saturated market for American cars and a shift in automated auto manufacturing to nations outside North America. Worldwide investment in industrial robotics had surged between 2003 and 2005 with more than 52,000 robots sold in Asia (including Australia and New Zealand). Predictably, the greatest sales in Asia during that period were in Japan, where 37,000 units were installed. However, the market for industrial robots in other Asian nations increased significantly. For instance, installations in the Republic of Korea rose by 17 percent, while those in China, Malaysia, the Philippines, Indonesia, Singapore, and India grew by approximately 125 percent. Orders from nonautomotive areas including the life sciences, pharmaceutical, and biomedical industries jumped 30 percent in 2005 and are expected to continue to increase significantly by 2008.

Automotive manufacturers and parts producers have also been the major users of robots in Europe, although other markets are also expanding there. For instance, sales of robots to the European machinery industry increased by 22 percent, and installations in the chemical industry rose by 72 percent. According to the joint survey, during this period more robots were installed in the food industry in Europe than in the Asia and America.

Taking into consideration robot attrition and replacement, there are now estimated to be over a million robots in use worldwide. Early in 2006, the RIA estimated that around 160,000 of them are installed in American manufacturing operations alone, making the United States currently second only to Japan in the use of industrial robots. It is notable that while the United States has been a leader in research and development and sales, it is at the bottom of the list when it comes to robot density, that is, number of robots per worker in an industry. For example, in Japan there were an estimated 329 robot installations per 10,000 workers in manufacturing in 2004. During the same period, Italy had a robot density of 123 and Sweden 107. Finland and Spain followed with 86 and 81 respectively, and France with 78. The United States was last with only 69 robots per 10,000 workers in an industry. It is not possible to unequivocally attribute this to any one factor, although competition from global neighbors, differences in attitudes about labor, availability of a workforce, and attitudes about investment in robotics for long-term growth are certainly influences.

EXPANSION

In the early twenty-first century, shifting political, economic, and social climates motivated governments and businesses to increase their investment in robots. Advances in microprocessing, along with reduced cost of parts like actuators and sensors motivated more nations to invest in robot-assisted manufacturing. This in turn has created a niche for multipurpose, small and micro parts assembly robots. For example, in 2006 the Spanish company Fatronik in collaboration with Lirmm (Laboratoire d'Informatique de Robotique et de Microélectronique de Montpellier) introduced its QUICKPLACER, a high-performance industrial robot that it calls the most rapid handling robot in the world.

Its four actuators are coordinated to enable the 4-degree of freedom (DOF) manipulators pick up and position over 200 items weighing up to 2 kg (4.41 lbs) per minute. Fatronik boasts that the robot's high acceleration and braking capacity is five times that of a Formula 1 racing car, allowing for a twenty per cent increase in productivity over its competitors. Precision manipulators like QUICKPLACER can be applied to a wide range of applications that includes food packaging, quality control of agricultural goods, packaging health and beauty aids, and electronic component assembly.

A more competitive atmosphere has led to diversification, acquisitions, and mergers linking robot manufacturers with producers of important ancillary systems. For instance, in 2000 Adept Technology Inc. of San Jose, California, a successful producer of SCARA industrial robots since the 1980s acquired a number of semiconductor and fiber optic operations to position itself as a leader in those markets. Adept also produces motion controllers, which it sells to other robot manufacturers like Stäubli. In 2006, Stäubli itself announced a partnering agreement with the Natick, Massachusetts-based Cognex® Corporation, the world's leading supplier of

QUICKPLACER high-performance industrial robot. © Fatronik. Used by permission.

machine vision systems. The agreement helps Stäubli remain a leader in robotics, since vision systems are so crucial to robot accuracy. Indeed, innovations in vision systems are at the heart of the expansion and diversification of robotics.

Vision Systems

In the 1960s, computer scientists used symbolic representations of objects to explain vital issues like how the edges of objects come together (see Chapter 4). Subsequent vision systems developed to help industrial manipulators identify and distinguish parts on an assembly line use algorithms to interpret the visual input from cameras mounted on or near the manipulator. Service robots require more precise and reliable vision systems to enable them to distinguish objects from people from different angles. Despite improvements in laser scanners and pattern recognition software, robots have difficulty with elements like texture, ambiguous shapes, and unexpected movement in their range of view, particularly while they are moving. A great deal of research has been devoted to solving these problems.

In 1996, a research team at Oklahoma State University designed a fuzzy neural net system to compensate for ambiguous data collected from a robot's cameras. Their FUZMAP was tested on a silverware sorter for a commercial dishwashing system, and could distinguish highly reflective or specular surfaces of materials such as metal, ceramic, or stone. Work has also been done with color recognition and laser stripe scanning to enable robots to identify previously unknown objects under normal ambient lighting conditions, or where there is interference from reflective objects, or people crossing its visual field. Some researchers have tried to develop mechanical analogies of the human vision system. For instance, in 2003 computer scientists from the University of Maryland at College Park demonstrated their ARGUS EYE, named after a god from Greek mythology that had eyes all over his body. This omni-directional robot vision system is a more advanced approach to the 3-D vision system Hans Moravec developed for the STANFORD CART (see Chapter 5). When we look around us, our eyes move to take in a lot of information, which our brain processes as a comprehensive picture. To simulate this process, Moravec used one camera that he had to reposition nine different ways to take in various aspects of the visual field. The ARGUS EYE accomplishes this with nine individual cameras attached to a frame that send raw data to its computer to build a comprehensive model of the environment.

Object recognition software forms the basis of PENELOPE, a two-armed robotic scrub nurse recently introduced by Robotic Surgical Tech Inc. to free up its human counterpart for patient care in the operating room. PENELOPE, which was first used at Columbia Presbyterian Medical Center in New York in 2005, can correctly distinguish a number of similarly shaped surgical instruments, select and hand them to the surgeon on request, and retrieve used instruments. PENELOPE compares the changing field during the operation with a base image of the instrument array taken beforehand. It can distinguish nonoverlapping objects by measuring such aspects as instrument length, width, and shape, size and angle of the instrument tip, and finger hole openings, and by comparing them from the background using color differentiation.

Evolution Robotics' Visual Pattern Recognition (ViPR^{TM}) system can recognize multiple objects simultaneously in real time across a wide range of angles, while compensating for distortion and varied lighting conditions. Its 1.4GHz processor can process at least 15 images per second, at a 208 by 160 pixel resolution. In January 2006, Evolution debuted their LaneHawk visual scanning system, a loss-prevention solution for retailers that identifies objects by shape and packaging rather than by barcode. LaneHawk uses the award-winning ViPR^{TM} technology to identify innocent or deliberate placement of goods in the bottom of the shopping basket or on the under rail of the cart. The technology is currently in use in over 100,000 machines worldwide including robots.

ViPR^{TM} is complemented by Evolution's low-cost NorthStar positional awareness navigation system, which uses triangulation to measure position and heading from a device projected onto a ceiling or other visible surface, and which updates 10 times per second. WowWee Robotics, maker of ROBOSAPIEN and ROBORAPTOR, announced in March 2006 that in an effort to offer customers truly autonomous and intelligent mass-market robots, it is integrating Evolution's ViPR^{TM} and NorthStar systems into their entertainment robots. Evolution also offers Visual Simultaneous Localization and Mapping (vSLAM®) for mobile robots, which uses a low-cost camera and dead reckoning to measure the environment. vSLAM® builds a map of its path using the captured landmarks, enabling the device to navigate smoothly using the preexisting map, or building a new one on the run, adding the new information to its knowledge base.

The most sophisticated vision systems still cannot give robots all the information they need, for instance, consistently and reliably distinguishing one person from another from behind, or under poor lighting conditions. Consequently, some engineers are developing methods of sensorial compensation to overcome the limits of robot vision. For instance, the M4 mobile robot project at MIT has incorporated a thermal imaging camera for night vision and a complex of visual and cognitive solutions to provide the robot with texture segmentation, gaze inference, and object 3-D reconstruction. Other research has focused on *haptics*, or the science of touch.

Haptics

Industrial robot manufacturers and users have always been concerned with the accurate grasping, handling, and passing of objects of different sizes, shapes, densities, and materials on assembly lines. Tactile sensing is even more crucial in the case of service robots meant to work in close proximity to humans. Some researchers argue that it is more critical than vision, since humans can survive without the sense of sight. In situations where our vision is inadequate or where there is not enough light to distinguish objects, we rely on touch to make such distinctions as solidity, density, sharpness, weight, even moisture level and temperature that are crucial for our navigating the world and staying alive.

The expansion of research in prosthetics for humans, domestic robots, and especially humanoid robots has motivated funding for more haptics labs, including the one founded by Alison Okimura at Johns Hopkins University in 2000, and at Princeton University where recently electrical engineers Stephanie Lacour and Sigurd Wagner experimented with nanosized flexible sensors embedded in a silicon-based synthetic skin. A joint research effort by the Intelligent Robotics and Communication Labs, ATR, Kyoto, and the department of Adaptive Machine Systems, Osaka University, developed soft, tactile sensor-embedded skin to cover the entire bodies of their humanoid robot prototypes. The experiment, led by Takahiro Miyashita, is meant to develop an accurate way for robots to identify such aspects as posture and position using touch when their vision systems cannot provide adequate information about the humans with whom they are interacting.

In 2005, Vladimir Lumelsky of NASA's Goddard Space Flight Center Vision for Space Exploration lab began developing a flexible skin embedded with over 1,000 infrared sensors that according to Lumelsky will enable robots to work effectively alongside humans in space. Similarly, Japanese researchers at the University of Tokyo are developing semiconductor temperature sensors and electronic pressure sensors embedded into a flexible plastic skin that enable robots to make multiple judgments about objects they encounter. The team expects to expand the sensing capabilities to include detection of sound, light, humidity, and strain.

Dexterity

Engineering reports often state that the development of highly dexterous humanoid hands with many DOF will make it possible to supplant

The Shadow Robot Hand with forearm. © Copyright Shadow Robot Company Ltd. Used by permission. contact@shadow.org.uk

human labor with robots in many more situations. During the last 15 years, several three, four, and five-fingered hands have been developed at universities including Waseda and GIFU in Japan, and Columbia and MIT in the United States, as well as at NASA, and independent robot companies such as the Shadow Robot Company in the United Kingdom. Shadow has commercialized the impressive five-fingered DEXTEROUS HAND that they claim reproduces as closely as possible the 24 DOF of the human hand. It has 3 DOF and four joints in each finger, and 5 DOF and five joints in the thumb. All of the measurements for its design were taken directly from the human engineers in the lab. They produced a self-contained system of metal and plastic, with all muscles and the valve manifold in the forearm, providing force output and movement sensitivity comparable to the human hand.

The hands for NASA's ROBONAUT were designed with the unique requirements of space operation in mind. The legless, teleoperated humanoid is being developed to stand in for astronauts in dangerous extravehicular activities (EVAs) on the International Space Station (ISS) or shuttle flights. All of its parts must be designed to withstand the extreme temperature variations of space, meet or exceed out gassing restrictions to avoid

NASA/JPL ROBONAUT demonstrating hand dexterity. Courtesy NASA/JPL-Caltech.

contamination of other space systems, and achieve long user life in a vacuum. The hands are the necessary size and strength to meet maximum EVA crew requirements, and joint movement meets or exceeds the human hand in a pressurized glove. ROBONAUT's ability to work with instruments as small as tweezers is a result of its 2 DOF wrists and 12 DOF hands. Finger movement for each hand is divided between a dexterous set comprised of two 3 DOF fingers (pointer and index) and a 3 DOF opposable thumb, and a grasping set that includes two 1 DOF fingers (ring and pinkie) and a palm DOF. ROBONAUT has not yet been used in a space mission, but it has tested well for precision EVA activities.

ROBOT LEARNING AND SOCIALIZATION

Unlike industrial robots that are trained to perform a pattern of movements within a restricted space, service robots must be equipped to deal with unexpected encounters with animate beings and inanimate objects in unconstrained environments. The learning curve for behavior-based service robots like the ROOMBA vacuum is not very acute, since it requires relatively simple coordination of their sensors and actuators to keep them from bumping into walls or animate beings, or falling down stairs.

However, the larger, more complex robots being designed for institutional and domestic environments require protocols for safe and compatible interaction with users who are not engineers. During the last 15 years several labs that have devoted resources to providing robots both the physical dexterity and social skills that will make humans comfortable interacting with them have been influenced by the behavioral and cognitive sciences. For instance, engineers at the Human/Robot Interaction Lab at Tsukuba, Japan, turned to Piaget's studies in early childhood learning as a rationale for teaching their humanoid robots through imitation.

Face Robots

A series of explosions resulting from human error in Japanese chemical plants in the 1970s motivated research to improve communication in the workplace. The issue of worker safety, along with the Japanese commitment to produce autonomous humanoid service robots, inspired experiments with robotic faces based on studies in the relationship between human facial expression and understanding. Fumio Hara, a computer scientist at the Science University of Tokyo began such a project in the mid-1990s. By 1996, they had reported that their prototype could recognize human facial expressions an average of 85 percent of the time (compared to 87 percent recognition by other humans), and could express such emotions as happy, sad, or angry through the use of its 18 air-pressure driven microactuators. Part of their ongoing work included the addition of synthetic skin and human features to give the robots a look that would be pleasing to humans.

Dr. Cynthia Breazeal was working on COG at MIT in the late 1990s when she was inspired to begin KISMET, a sociable face robot project for her PhD project. The project is inspired by the relationship between an infant and its caregivers, and so KISMET gives the appearance of being an infant or toddler. The mechanical component of its visual attention system includes microphones for auditory input, two foveal cameras behind its eyeballs, and two wide-angle cameras behind the "nose," and simplified motorized eyebrows, lips, ears, and an articulated jaw. KISMET also has a 3-axis neck and a gyroscope in its head (simulating our vestibular-ocular reflex). A 15-computer array works in a distributed manner to control movement, voice, and attention. The robot pays attention to either moving things, skin-colored things, or those with saturated colors. Breazeal and her team worked with KISMET using the same brightly colored toys found in any nursery, and the kind of language and gestures people use with young children. The robot's internal drives control change in mood. For instance, if it has been looking at a toy with saturated color too long, it might get bored, and be more attentive to someone moving across the lab. KISMET interacts with visitors using its auditory and speech functions to respond to *prosody* or pitch variations in a person's voice, to which it may express moods such as happiness, anger, surprise, or fear. It is important to understand that KISMET is not a child and does not really understand what it is doing. It has merely learned through demonstration and imitation to use its expressive capability in ways that are familiar and acceptable to humans. These experiments are valuable method of learning how to instill social behaviors in robots that will make humans comfortable with them, and enable them to function successfully.

Others are working on face robots to provide them with realistic coun-tenances. For instance, David Hanson of the University of Texas, Dallas, and an alumnus of the Rhode Island School of Design, developed a series of realistic android heads using lightweight facial hardware, which enables the robot to express 28 facial movements. They are covered with a synthetic skin he invented by combining an elastomer and foaming agent polymer. Hanson's Frubber® gives the elasticity and look of human skin without its complexity. While his hyper-realistic animatronic heads may useful to movie and theme park designers, Hanson has also demonstrated his K-BOT head at scientific conferences to interest researchers in using his inventions in areas like cognitive science and education.

Emotion and Intuition

Familiar features are not enough for robots to be able to communicate effectively with humans according to some researchers. Engineers at the Shuji Hashimoto lab at Waseda University argue that since humans often utilize both logical and conscious thinking in tandem with illogical or intuitive thinking—which they call "Kansei"—a robot that interacts closely with humans must also have both capabilities, which they see as equally important. Engineers working on the Waseda University Anthropomorphic Head-Eye Robot (WE-series) argue that a robot must be able to perceive the effect its touch and voice have on human users. Therefore, it needs a complex perception system supported by an effective natural language capability that will enable it to receive and express emotional information. In addition to facial features, a flexible neck, and a microphone for auditory input, they also equipped their robot with a force sensor to develop physical social awareness about the difference between such actions as touching, pushing, hitting, and stroking.

Researchers working on the COG project at MIT have taken different approaches to providing the BBR robot with the skills it needs to learn to interact with humans. For instance, Bryan Adams has observed that while most of the work in humanoid function has been with haptic, visual, and audio sensing, "a great deal of the information governing the organization and execution of limb movement comes from the energy metabolism that supplies muscles with energy" (Adams, 1–2). Consequently, his team has experimented with virtual pancreas/adrenal gland, liver, and adipose tissue (fat) modules to communicate to the robot the energy level in its limbs, and to provide the right amount of fuel necessary for the robot to move its arms. As the system recognizes the overexertion trigger built into the robot's muscle, it reduces the motor power, simulating human muscle fatigue. Thus the robot can "accurately judge the state of its arm" and react in a human-like manner to the environment. However, Adams also sees this physical control system as the basis for the robot's emotional model. He has suggested that by refining the biochemical processes model, the humanoid will be able to combine neurologically based emotions like loneliness or fear with biochemical states like hunger, fatigue, or excitement.

Brian Scassellati has argued that successful human–robot interaction will depend on robots learning social cues in the same way young children do. The basis of his work with COG draws on the work of such behavioral and cognitive theorists as Alan M. Leslie and Simon Baron-Cohen who have written about two principles, *Theory of Body*—the ability to distinguish animate from inanimate objects, and *Theory of Mind*—the ability to recognize that there are other individuals in the world with different perceptions and intentions from our own. These are thought to be critical steps in childhood development, and in particular to language acquisition, self-recognition, and perhaps even imaginative and creative play. Consequently, researchers like Scassellati are invoking them as necessary to the development of personal robots.

Researchers at Carnegie Mellon (CMU) agree that sociability is also important to the success of mobile service robotics. The many unspoken social rules humans adhere to in day-to-day activities inspired a social proximity project, in which a mobile robot called XAVIER was taught how to use social cues to do an errand, "going for coffee." They programmed into the robot such social patterns as the acceptable amount of space to leave between itself and the person in front, and going to the end of the line. These kinds of experiments help researchers to solve related perception problems like recognition of people from the side or behind so that they are not interpreted merely as obstacles. Many humanoid robots are being field-tested to learn more about human–robot interaction. For instance a humanoid called ACTROID welcomed visitors to the Robodex 2005 in Aichi, Japan, and gave directions and other information while maintaining a pleasant and helpful demeanor.

Speech Synthesis

Many researchers identify natural sounding speech as essential to comfortable human–robot interaction. Consequently language processing is an essential element in the design of service and entertainment robots. For instance, the speech component of surgical robots like PENELOPE allows the surgeon to conduct an operation in the same manner as she/he would if a human operating room nurse was assisting. Some researchers have argued that a robot's ability to understand what is said, and to ask and respond to questions may be a matter of life and death. For example, a robot sent into a disaster site may encounter survivors who cannot move, but who can verbally relay information about their state and location. The tone of a robot's voice is important for clarity and even trust or comfort in situations like these.

People have been trying to simulate human speech since antiquity. By the eighteenth century, mechanics were using components of musical instruments to simulate the human voice. In the nineteenth century Sir Charles Wheatstone, the man credited with the invention of the telegraph in England, improved upon an artificial vocal tract designed by Wolfgang von Kempelen in 1791. The device comprised a large bellows attached to an air chamber, a reed like those used in wind instruments attached at one end to a leather "resonator," a whistle, a minor bellows, and a series of levers. By manipulating the flow of air through the chamber using the whistle, resonator, and other components, Wheatstone was able to simulate vowel, consonant, and unvoiced sounds.

The first modern robot voices were the result of research in speech synthesis begun in the 1930s. Westinghouse demonstrated their progress in this area through their exhibition robots, mentioned in Chapter 3. During the same period, engineer Homer Dudley was attempting to invent a more efficient method of converting vocal signals in telephone transmissions for Bell Telephone Labs. His VODER, which was operated like a musical organ, first made headlines when it was demonstrated at the 1939 World's Fair. During World War II it was used as a voice encryption device. Dudley's invention was not widely adopted even though it did significantly reduce the bandwidth necessary to transmit speech in existing telephone systems. Nevertheless, it inspired further research in speech synthesis. Advances in computing technologies made possible systems that could convert electronic signals into robotic speech, such as the Kurzweil Reading Machine (KRM), a print-to-speech reader for the blind invented by Ray Kurzweil in 1975, and the automated "operators" that we encounter in computer-assisted phone calls.

The concept of building an artificial vocal tract has been revived in order to replace the stilted, computerized voices of robots with a more natural sound. For example, Hideyuki Sawada and Shuji Hashimoto of Waseda University are developing a sophisticated version of Wheatstone's Victorian-era machine with synthetic lung, windpipe, throat, and vocal cords. It listens to its own utterances and through its neural network it learns to modify the flow of air and shape of the vocal tract to produce more human-like sounds. Designers hoped to eventually add a tongue.

The sounds produced by the artificial vocal tract in development at Kagawa University in Japan are dependent upon the tension of a rubber vocal cord, the shape of the tract, and airflow velocity. In place of Wheatstone's bellows, a compressed air tank forces air into a plastic voice-box chamber. There, airflow makes the rubber vocal cords vibrate. These basic sounds are fed into a flexible tube that represents the human vocal tract. Just as the shape of our vocal tract alters to dampen out certain frequencies, the Kagawa system can be altered using motorized rams that push on the sides of the silicon tube. A number of other labs have contributed the initiative. For example Toyota, whose Partner humanoids can play horns, have synthetic lips that help them form the air pocket necessary to blow into the instruments.

LOCOMOTION

By 2000, there were humanoid prototypes in development in a dozen countries. A number of engineers had succeeded in getting their robots to walk independently on smooth ground or up stairs, although some of them still were dealing with challenges of dynamic balance, and wrestling with the reality of production cost outweighing functionality. Consequently, a number of labs reduced the size of their prototypes and began gearing them to the entertainment and exhibition market. Among the most visible were Honda's ASIMO and the smaller Sony Dream Robot (SDR). During the next few years, improvements were made in stability and speed and on December 18, 2003, Sony unveiled the latest iteration of its dream robot, now called QRIO, and demonstrated it running and jumping. Subsequently QRIOs have participated in the RoboCup soccer competition discussed

As part of the "ASIMO Technology Circuit University Tour" ASIMO demonstrates dance moves before a capacity crowd of more than 1,000 to kick off Robotics Week at Purdue University March 10–19, 2005. Courtesy Purdue News Service. Photo/Dave Umberger

below. Although early in 2006, Sony discontinued QRIO development in order to redirect capital to other consumer lines, it could boast that it had been the first to achieve stability in a robot while both feet are off the ground. In 2005, QRIO, ASIMO, and the newly unveiled Toyota Partner humanoids performed together for the first time at the Expo 2005 in Aichi, Japan. By that time, Honda had begun demonstrating ASIMO running.

Korean developer Robotis announced soon afterward that it had developed a low-cost solution to the expensive humanoids developed by Sony, Honda, and Toyota. For robots meant to work in homes and institutions, stable walking and running are a big advantage that will also be advantageous to researchers working to adapt humanoids to industrial and military applications.

Many of the wheeled or tracked robotic vehicles and legged robots described in this volume are either too costly or too massive to use in certain situations. Consequently a number of government-funded research

Humanoids like ASIMO promoted as friends and helpers recall GORT in *The Day the Earth Stood Still* (1951). Courtesy of Photofest

projects are devoted to alternative mobile robot configurations that would be useful for reconnaissance, search and rescue, and delivering supplies to soldiers or disaster victims in multiple terrain environments. Some of these, which will be described below, are simply smaller versions of tanks and armored cars that can be thrown into a building or driven by remote control into spaces too small for humans or other robots. Others are based on a theory developed in the 1980s called passive/dynamic walking.

Passive/Dynamic Walking

In the 1980s, roboticist Tad McGeer, working at Fraser University in British Columbia, produced some simple, completely passive robots that could walk down a ramp as efficiently as a human without the use of any expensive motors or controllers, propelled only by an initial nudge and gravity. The problem with McGeer's design was that the robots could only walk downhill, which limited their functionality. However his idea was provocative, and in the wake of extremely expensive and complex humanoids like ASIMO, some engineers have taken up the challenge of producing biped robots that maintain the spirit of McGeer's experiment but add some support mechanisms to enable the robots to walk on flat surfaces or up an incline, perhaps carrying weight. In 2005, researchers from Cornell, MIT, and Delft University of Technology demonstrated semipassive robots that can walk with relative stability on flat ground. Andy Ruina of Cornell University codesigned a walker that uses two small motors, some wiring at the hip, and two batteries to weigh it down. It is the most energy-efficient of the experiments, providing a bipedal gate with about the same energy use as a human walker, and at about 1 percent of the cost of producing an ASIMO. DENISE, built by Martijn Wisse of Delft University of Technology, uses slightly more energy, but performs gracefully with the addition of pneumatically powered devices mounted at the hips. The MIT prototype, equipped with a large onboard computer, is impressive but uses the most energy and is as expensive to power as the Honda humanoid. Critics observe that at this point, passive/dynamic walking does not allow for climbing stairs or carrying heavy loads, or even side-stepping obstacles. Still, many researchers are enthusiastic that further research will result in more functional, energy-efficient designs.

Other researchers have taken a more novel approach to robot movement that combines the best elements of legged and wheeled locomotion to yield a more robust, flexible, and cost-effective robot. Several different configurations have been produced, including Prototype Of Legged Rover (PROLERO), designed by Martin Alvarez at the European Space Agency in 1996, RHEX designed in 2000, and the Case Western WHEGS™, built in 2001. Each of these small BBR-type robots use small motors and some method of rotating the legs as if they are the spokes of a wheel.

Most recently, Professors Damian Lyons and Frank Hsu in the Robotics and Computer Vision Lab in the department of computer science at Fordham University in Bronx, New York, received a grant from the U.S. Department of Defense to develop a low-cost, lightweight robot that will be able to maintain stability while moving through changing terrains, from level to inclined, smooth to rocky, and soft to hard. Like the projects mentioned above, this project is meant to use the best features of legs and wheels. They hope to produce a robot that can be easily carried, dropped or thrown into a target area to deliver supplies or perform reconnaissance, and quickly recover to a standing position, and move with stability and efficiency over any terrain. Their ROTOPOD, which looks a bit like a camera tripod, can take a step with one leg, rotate the wheel part of the mechanism around the leg, and step down again. Thus the horizontal rotational movement of the center wheel is translated to the vertical legs, which operate like spokes. Depending on the aspect of the translational legs and the wheel mechanism, the robot will theoretically cover a projected area. The researchers are addressing such issues as how the robot will recover from falling over, or finding a long-lasting battery that does not overtax the robot's mass during movement.

Advances in robot perception, interaction, and locomotion made in research labs have been commercialized during the last decade for use in the health care, public relations, and domestic markets, as well as for use in exploration and combat.

MEDICAL

The health care industry was one of the earliest proponents of mobile service robots. Service robots had first entered hospitals in the 1980s when Joseph Engelberger founded HelpMate Robotics, discussed in Chapter 5. At a special U.S. Congressional hearing on the future and value of robots held in 1989, hospital representatives argued that using robots for tasks like delivering specimens or decontaminating surgical instruments would save many thousands of dollars per robot annually while displacing very few highly paid technical workers. Their appeal was in no small way related to the rising cost of running health-care facilities that has continued to result in the merging or closing of area hospitals, the loss of nursing and technical staff through layoffs and job disenchantment, and consequent reduction in people entering the nursing profession. Since that time, several other companies have emerged to supply mobile robots to hospitals around the world. TUG, produced by the Pittsburgh-based Aethon Company currently serves at Magee Women's Hospital in Pittsburgh, Pennsylvania, making deliveries. Like HelpMate, TUGs are equipped with sophisticated vision sensors. They use wireless radio signals for signaling elevators and opening electric doors, and are programmed with an internal world model of the hospital. ROBOCART, produced by California Computer Research, uses a fixed path routing system that is reminiscent of Schmidt's version of the

Stanford CART discussed in Chapter 5. It is being used at the Mayo Clinic and elsewhere for transporting blood samples. A tape is laid on the floor along the route that ROBOCART is to follow, and a sonar array used to judge its distance from objects is used for obstacle avoidance. While this seems primitive compared to other systems, it is also a less costly alternative for repetitive tasks along the same route.

Japanese roboticists, who have their sights set on humanoid caregivers, demonstrated a prototype android NURSEBOT at Robodex trade show in 2003. Many Japanese researchers agree that humanoid robots provide both the dexterity and a comforting personal association that nonhumanoid machines cannot. There is currently no humanoid robot that is ready to replace a health-care worker; however, nonhumanoid and semihumanoid mobile robotic assistants are becoming more prevalent.

CMU's *Robots and People Project* team has been experimenting with friendly mobile robots that can assist elderly or recuperating patients in their home in order to shorten or avoid institutionalization. Their "nursebot" PEARL is essentially a wheeled mobile robot platform with a touch-screen computer for a user interface, and a platform that can be used to transport items or as a seat. It is not humanoid in form, but since the designers share the view of many researchers that face and voice are the key to communication, PEARL has an expressive face, and can communicate through speech as well as the graphical interface. Currently the designers are field-testing PEARL using different facial features and voices to study which ones will be the most attractive to users.

Gecko Systems' Mobile Service Robot (MSR) line, targeted to health care and personal assistance, has solved several problems associated with mobile robots. For instance, the 4-foot-tall MSR 3.4's aluminum GeckoFrame^{TM} helps to keep the unit's total weight at just over 100 pounds including its 50-pound battery. Its light weight and efficient use of battery power allow the unit to track its client for up to 24 hours without recharging, or transport 100 pounds above its own weight for over 12 hours before it requires recharging. Gecko has integrated a series of proprietary perception and communication systems into its MSRs: GeckoChat^{TM} verbal interaction package that is capable of monologues and interactive dialogues, as well as verbal control of GeckoNav^{TM}, the automatic self-navigation software that gives the robot autonomy, and GeckoTrak^{TM} software that enables the robot to follow its client around. Gecko designed its MSR with sociability in mind. The robot can be programmed with familiar, comforting or humorous language, which can be adjusted for tone using its equalizer. The unit sells for around $20,000 and includes travel, meals, and lodging for three days of on-site user training.

A promising market for medical robotics drew the interest of university-based researchers. In 1993 the CMU robotics department formed the Center for Medical Robotics and Computer Assisted Surgery. Researchers from CMU, MIT, and Johns Hopkins established the NSF-funded Engineering Research Center for Computer-Integrated Surgical Systems and Technology (MR CAS) under the direction of CMU Professor Takeo Kanade in 1998. In 2001, CMU established the Medical Robotics and Information Technology Center (MERIT) to create new robotic technologies for the health-care industry. Among MERIT's accomplishments is HipNav, a computer-aided system that helps surgeons accurately place implants during replacement surgery. Professor Kanade and orthopedic surgeon Anthony Di Gioia, who has since performed hundreds of hip replacements using HipNav, developed the device.

In the 1990s, a number of medical professionals teamed with engineers and business experts in medical robot startup companies. For instance, orthopedic surgeon William Bargar and researcher Howard Paul developed a robotic arm for surgical procedures called ROBODOC. They demonstrated its efficacy by using it to perform a hip replacement on a dog. Howard Paul formed Integrated Surgical Systems to commercialize the device; which received FDA approval after it was successfully used in a human hip replacement in 1992.

Surgeon Fred Moll, electrical and mechanical engineer Rob Younge, and MBA John Freund formed Intuitive Surgical in Sunnyvale, California, in the 1990s to design and market surgical equipment. In 1998, they introduced the DA VINCI® surgical system. Although it is not currently marketed for remote location surgeries, its InSite® vision system provides the surgeon with an actual (rather than virtual) high-resolution 3-D visual of the surgical field, and the ability to control the manipulators from a comfortable surgeon's console. The patient side cart incorporates industrial robot manipulator technology for its design. Its four arms, each with 7 DOF and wrist manipulators, support a full range of use-specific instrument attachments that are controlled by the surgeon turning the remote dials intuitively (in the same direction as the intended movement). However, DA VINCI is not technically a robot since it makes no programmed or autonomous movements. In fact one of its selling points is that it gives the surgeon more control. The U.S. Food and Drug Administration (FDA) has recently approved the $1.5 million machine to support a large number of minimally invasive surgical (MIS) applications including mitral valve repair, radical prostatectomy, gastric bypass, and cholecystectomy. DA VINCI® is now in use in over 300 academic and community hospitals in at least 19 countries.

Several other manufacturers now produce robotic manipulators for the medical field including two German companies: OrtoMaquet, which produces the Computer-Assisted Surgical Planning and Robotics (CASPAR) device—an industrial robot, milling tool, and calibration unit mounted on a mobile base that is used to help a surgeon in orthopedic procedures like hip surgery, and Carl Zeiss, which produces a medical manipulator (MKM) that consists of a computer-controlled 6 DOF robotic arm and graphical programming and workstation.

PERSONAL ASSISTANTS

A growing number of service robots are being designed to be useful in different settings where humans need information or assistance. In addition to hospitals, they are being tested in convention centers, hotels, and museums. Between 1998 and 2002, Illah R. Nourbakhsh, Clay Kunz, and Thomas Willeke of the CMU Robotics Institute ran a mobile robot experiment at the Carnegie Museum of Natural History in Pittsburgh, Pennsylvania. The CHIPS project, initially organized to install a socially interactive, autonomous mobile robot at the museum, was expanded under the charter Mobot Inc. to include three more mobile robots. The experimental robots functioned as tour guides in different areas of the museum every day for the period of the experiment. While they achieved the goals of autonomy, educational function, and social interactivity, the researchers learned in the end that the $10,000 annual cost of maintaining a $200 thousand dollar machine would be prohibitive to museums.

Researchers have continued to develop mobile assistance robots with a view to lowering manufacturing and maintenance costs. The museum setting is still a popular testbed for human–robot interaction. For instance, in the spring of 2005, the London Aquarium installed three mobile semihumanoid tour guides designed by researchers at the University of Essex who wanted to test how humans relate to robots. Their goal is for the robots to be able to locate and help lost children and entertain people waiting in line.

In 1998, Katja Severin and Hans Nopper designed the first mobile personal assistant robot prototype for the German robotics company Fraunhofer IPA. Their CARE-O-BOTs are mobile platforms with a moveable interactive touch screen (not to be confused with Gecko Systems CareBots). They are designed with self-start and diagnostic functions, making them attractive to untrained personnel. They navigate autonomously, using a laser scanner for obstacle detection, navigation, and self-localization, but can be controlled using a joystick on the robot console, and monitored remotely over a radio Ethernet system. In the spring of 2000, three CARE-O-BOTs were installed in the Museum für Kommunikation in Berlin, where they successfully navigated through the environment interacting with visitors. Configured as three different characters, they provided different functions, one as a greeter, one as a guide, and one as pure entertainment.

CARE-O-BOT II, equipped with a manipulator arm, adjustable walking supporters, a tilting sensor head, and a hand-held control panel was unveiled at Hannover Messe 2002 automation exhibition, where it exchanged business cards with visitors. The robot's laser sensors and other perception software allowed it to detect visitors, whom it approached and offered a business card with its manipulator arm. The motion of the arm was coordinated with speech output and graphical information on the screen to enhance intuitive interaction with humans. CARE-O-BOT II responds to user requests by either voice or touch screen. Its producers say that the robot is always in the control of the user, who can interrupt the robot's movements and modify its program. CARE-O-BOTs I and II, already in service, are marketed to provide the elderly some measure of independence while still offering services like reminders to take medication, or the location of rooms in an institution. A good deal of government-funded research has been provided to develop robotic solutions for planetary exploration, hazard detection and cleanup, reconnaissance, and combat.

EXPLORATION

In the 1990s NASA planned missions to Mars that would employ these small lightweight mobile robots to demonstrate the economy and safety of using telepresence robots for space exploration. The Jet Propulsion Laboratory's (JPL) Mars rover prototypes Rocky I–IV were based on Colin Angle's TOOTH (See Chapter 5). Meanwhile, MARSOKHOD, a joint mission of the International Mechanics Group run by the U.S. and Russian space agencies tested wheeled rovers in a number of remote Earth locations that are similar to the Martian surface: the Kamchatka Peninsula, Siberia (1993), the Mojave Desert in the United States (1994), and Kiluae volcano, Hawaii (1995).

NOMAD, a four-wheeled rover developed at CMU was sent on meteor-finding missions between 1997 and 2000 to environments that have much in common with the Martian terrain. In 1997 it completed a 40-day, 131-mile exploration of the Atacama Desert in Chile, while being teleoperated from Pittsburgh and California. Between 1998 and 2000 it performed three other expeditions in Antarctica. The success of the finding missions

SOJOURNER Martian Rover. Courtesy of NASA

was in part due to NOMAD's vision system: two CD cameras placed at eye width on its mast to provide depth perception, and a panoramic camera fitted with a convex optical mirror that together provided remote operators with a clear 360-dregree view of the field. A laser rangefinder enabled obstacle detection while orientation sensors detected potentially hazardous terrain. NOMAD detected rocks in the ice and snow using a high-resolution color camera, and performed analysis using a reflection spectrometer mounted on its manipulator that could get as close as 1 centimeter of a rock face.

On July 4, 1997, PATHFINDER landed in the Ares Vallis region. There it dispatched the microrover SOJOURNER.

The 22-pound robot was 10.9 inches high, 24.5 inches long, and 18.7 inches wide, and comprised a low conductivity Warm Electronics Box (WEB) covered in gold foil. The heated container protected all of SOJOURNER's vital electronic components from the extreme temperatures of Mars. Equipped with a solar panel array of over 200 solar cells, the rover's onboard power output was 16 Watts, and backed up with a lithium battery for operation in low-light conditions. SOJOURNER moved on six 5-inch-wide wheels each equipped with cleats and an independent motor. The front and rear wheels had individual steering motors, which allowed the vehicle to turn in place. Its *Rocker-Bogie* suspension enables the rover to climb over rocks or other obstacles up to twice the wheel's diameter while keeping all six wheels on the ground, and to withstand a tilt of up to 45 degrees in any direction without tipping over. (However, it was programmed to back off if it encountered gradient angles above 26°.) PATHFINDER made atmospheric and meteorological observations; SOJOURNER utilized an alpha-proton X-ray spectrometer to analyze the composition of rocks using its sixth wheel, which was designed to lock in place to perform as an abrasion tool. Its wear characteristics indicated the composition of the samples.

The design of the SOJOURNER mission illustrates the goal of mutual cooperation between humans and robots in space. Both PATHFINDER and SOJOURNER were equipped with stereovision systems to return visual information from their position on Mars, and to assist teleoperation from Earth. The Lander relayed data gathered by the rover back to Earth. SOJOURNER used a laser range finder to locate goals, and was equipped with onboard AI to avoid accidents with the equipment during communications delays that could run as long as 41 minutes. It would attempt to cross obstacles up to three times, at which point it would radio an S.O.S. with stereo images to PATHFINDER, which relayed them back to Earth. The image data was used to construct the digital terrain maps the scientists used to instruct the robot how to get out of its predicament.

SOJOURNER covered only about 387 feet of the Martian surface, but when it stopped sending back images and data after 83 days, it had relayed 550 images and 15 soil analyses. The PATHFINDER returned about 16,000 images, some of which suggested that Ares Villas had once been a flood plane. Evidence of dried waterbeds at Gusev crater and the Meridiani plain motivated NASA to launch the Global Surveyor probe in 1999 to map all of Mars, and to send another rover mission. After two failed mission attempts, the rover SPIRIT successfully landed near Gusev Crater on January 3, 2004, followed by OPPORTUNITY, which landed on the Meridiani Planum on January 24. Once again the rovers and the scientists on Earth exchanged visual data, which was used to send the rovers further instructions. SPIRIT and OPPORTUNITY continue to contribute valuable information to space researchers. In July 2006, visual and rock

SPIRIT Martian Rover, dispatched from Pathfinder lander, July 4, 2004. Courtesy of NASA

sample analysis helped confirm that the ridges in the Martian surface were indeed caused by water wear and not wind.

Robots also play important role in the long-term cooperative development of the International Space Station (ISS). The Space Station Remote Manipulator System (SSRMS) by MD Robotics of Canada was successfully launched in 2001 to complete assembly of the ISS. CANADARM 2 (installed 2001) with 7 DOF, including a three-jointed shoulder, one-jointed elbow, and three-jointed wrist can change its configuration without moving its hands. Full joint rotation provides a wider range of motion than a human arm. It is coordinated with the Mobile Service Base System (installed 2002) in such a way that CANADARM 2 can travel the entire length of ISS, moving end-over-end to reach various locations.

WORKING TOGETHER—DISTRIBUTED ROBOTICS

In the 1990s a research team in Finland turned to biological societies as models for developing robots for use in continuous, long-term, autonomous operation in hazardous environments. Despite the lower intelligence of the individuals in societies of bees, ants, and other insects, the group acting together exhibits a higher intelligence and in this way survives in its environmental niche. The team adapted this strategy to a robotic "society" of multiple small robots with low-level intelligence that work together to complete a task. The individuals communicate only locally, on a member-to-member basis. The base station serves as the interface with the users, who monitor the operation through a graphical interface. The number of component robots can be adjusted to suit individual project needs. Meanwhile the high level of redundancy in the society makes the continuation of work possible even if one of the robots breaks down.

Search and Rescue

During the last 10 years, distributed robotics has been adopted for reconnaissance and search and rescue operations. For instance, the Swarmbot project, led by Francesco Mandada at the Swiss Engineering University, allows multiple small cylindrical mobile robots to operate independently, and then hook together for search and rescue missions. A team of SRI researchers began a DARPA-funded distributed robot project in 2002 with the objective of sending a cohort of robots into inhospitable environments like collapsed, burning, or smoke-filled buildings, chemical-spill sites, or terrorist-occupied structures.. In January 2004 the SRI team successfully completed an experiment using one hundred small autonomous mobile robots comprising 97 ActivMedia Amigobots and six ActivMedia Pioneer 2 AT mobile robots. The CENTIBOTs demonstrated that multiple robots working cooperatively could map, track, and guard objects or an area in a coherent fashion over a 24-hour period.

After the bombing of the World Trade Center on September 11, 2001, Robyn Murphy, professor of cognitive science and robotics at the University of South Florida, brought a team to New York to help search the debris at Ground Zero. They deployed several tiny robots that look like military tanks sans the artillery weaponry. Each "marsupial" carries onboard smaller "child" bots that are released to search in otherwise inaccessible areas.

At Ground Zero, the components communicated with each other through a wireless network, and consequently Murphy's team was able to locate five survivors. Other robots used at Ground Zero included shape-changing recognizance robots brought by a Colorado team led by Colonel (retired) John Blitch, and commercial mini pipe inspection robots produced

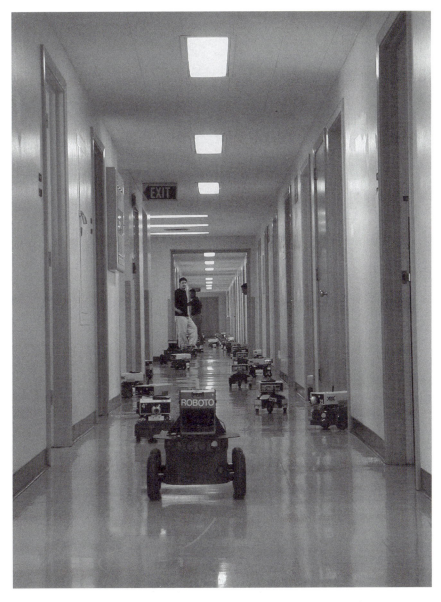

CENTIBOTS mapping demonstration. Courtesy of SRI International, California

by the British Columbia company, Inuktun. The company makes several different search and rescue robots, including an unmanned helicopter, tiny track vehicles, and the giant ENRYU, a track vehicle with cameras and arms designed to clear debris from disaster sights.

Combat

Increased terrorist activity during the last quarter century provided the rationale for funding a number of mobile robotics projects. For instance, although the Segway® Human Transporter (HT), released in 2002 did not revolutionize urban transportation as its designers expected, DARPA quickly purchased a fleet of them and distributed the versatile platforms to robotics researchers to study how the relatively inexpensive tool could be used in military situations. Now known as the SEGWAY® Robotic Mobility Platform (RMP) it is promoted for its durability, simplicity, and ability to run on a variety of power sources as a customizable unit for transporting payloads up to 181 kilograms (400 pounds). The Massachusetts-based iRobot Corporation, which receives a significant portion of its funding from DARPA, produces the PackBot, a small track vehicle that operated successfully in reconnaissance missions in Iraq and Afghanistan.

DARPA is also funding research on articulated robots for combat. The ROBO URBANUS project, a humanoid for warfare and reconnaissance proposed by a team at NASA's JPL under the direction of Adrian Stoica. They argue that the humanoid configuration is more suitable environments built for humans, and that such robots would inspire a higher level of acceptance than others from civilians in a war zone, while keeping human soldiers out of harm's way. They theorize that through a learning method they call "teaching/fostering," the robot would build on built on primitive needs, instincts, and reflexive behaviors, developing imaginative and intuitive skills needed to operate in a self-aware way in the field. An android with perceptual and motor skills similar to a human's could perform a wide variety of tasks like climbing, lifting, and carrying. The team is using a small HOAP humanoid made by Fujitsu for research purposes. However the team's long-term goal is to produce a full-sized humanoid that could be sent into combat and other hazard situations like scouting missions and rescuing soldiers and civilians from compromised areas. The U.S.-based JPL is only one of several agencies that are working on biologically-based robots with such capabilities. For instance, in 2005, the *Korea Times* reported that defense and communications technicians in the Republic of South Korea would begin a multibillion-dollar government-funded project to build armed, horse-like robots for combat, and Ikuo Mizuuchi demonstrated a humanoid called KOTARO that can climb a tree, bend to avoid branches, and pick up objects at the 2005 World Expo in Aichi, Japan.

In April, 2001 the GLOBAL HAWK robotic spy plane charted its own course over a distance of 13,000 km (8,000 miles) between California,

United States, and Southern Australia. Dozens of countries are now developing their own unmanned aerial vehicles (UAV). For instance, in 2005, the New Zealand-based TGR Helicorp, a commercial helicopter company, debuted their SNARK, a UAV for combat. It is touted for its lightweight Kevlar and carbon fiber construction, which contributes to its speed capability of 280 kmph, its quiet operation, its capability to remain airborne for 24 hours at a time. The SNARK runs on diesel fuel, making it easy to adapt to military use.

A number of commercial companies are receiving government funding to develop military support robots. Boston Dynamics, which makes virtual combat software for soldier training DI Guy, developed BIG DOG with funding from DARPA. Advertised as the "Most Advanced Quadruped on Earth" (bostondynamics.com), the dog-sized quadruped robot can walk, run, climb over rough terrain, and carry heavy loads. Its onboard sensors both help it navigate and monitor its own temperature, oil, and power levels.

ROVs

Undersea vehicles have been in use for decades, but a number of robots have recently been configured for undersea use. For instance, in 2000, a Triton XL marine ROV was used to prevent seepage from the sinking oil tanker Erika. A design team from Draper Laboratory led by Dr. Jamie Anderson recently developed a robotic tuna to perform underwater surveillance and rescue missions in high-risk areas such as mine-filled, radiation-compromised, or extremely low-visibility areas. They based their design of the Vorcity Control Unmanned Undersea Vehicle (VCUUV) on their study of a yellow fin tuna caught off the Long Island Sound. The robot's sensors, batteries, and motors are contained in its waterproof carbon fiber hull. Four independent hydraulic links are covered by flexible laminate "scales" that bridge the gap between the ribs allowing for smooth movement when the tail bends. The tail itself is covered in neoprene to reduce drag during swimming. In field tests it could dive up to 100 feet and swim for up to three hours at 24 knots.

PIPE CLEANERS

In addition to Inuktun, a number of companies develop robots to inspect and clean sewer and water pipes. During the late 1990s, as part of its initiative to develop robots for inhospitable environmental conditions, RedZone

NURP1 low-cost ROV can be deployed from small boats and dive to 1,000 feet. OAR/National Undersea Research Program (NURP), University of Connecticut.

Robotics Inc. produced HOUDINI, a prototype for a wheel-driven waste extrication system that could use handling tools like backhoe and plow to clear environmentally compromised sludge from pipes, tanks, mines, and caves. The Houston Texas-based itRobotics, founded in 2002, introduced a snake-like autonomous in-line pipe inspection robot as a cheaper and easier solution for inspecting closed, tubular objects as narrow as 6 centimeters in diameter. Their train-like robot pulls multiple carts equipped with sensors that indicate chemical changes like magnetic flux that may indicate cracks or wall-thinning. Utilizing an onboard computer and a proprietary locomotion system it can travel tether-free for kilometers, detecting flaws in pipes that may not be apparent from the outside. The company plans to build models for use in refineries, chemical and desalinization plants.

Another "snakebot" was developed in 2005 at the University of Michigan College of Engineering under the direction of Professor Johann Borenstein, head of the mobile robotics lab. The 26-pound OmniTread whose movement resembles a high-tech slinky as it climbs pipes and stairs, rolls over rough terrain and spans wide gaps to reach the other side. Its designers credit its unique tread design to OmniTread's ability to avoid stalling on rough terrain.

The tetherless OmniTread OT-4 with onboard power for up to 75 minutes of operation here shown in a pipe inspection demonstration. © University of Michigan, 2006. Used with permission.

PERSONAL USE

The most widely used domestic robots are small, low-cost cleaning robots. Among the leaders have been automatic cleaners that wander the walls and floors of swimming pools removing algae and small debris, and floor vacuums like iRobot's BBR disc ROOMBA and their recently released SCOOBA shampooer. Several companies including Electrolux have released more expensive robotic vacuums that cost over $1000 to do what ROOMBA does for around $300. In general, however, consumers have warmed to the idea of leaving a small, relatively nonintrusive robot wandering one's home picking up the never-ending accumulation of dust and small debris with little more trouble to the user than emptying its receptacle. ROOMBA has been improved since it reached 1 million sales. The ROOMBA DISCOVERY model is equipped with a 16-BIT microprocessor and flash memory, and can now adjust its movement pattern to pick up dirt more efficiently, and detect when it is running low on power and look for its recharging station.

By the late 1990s, several companies were market-testing personal "servant" robots. For instance, Sharper Image sold a 25-inch-tall, 24-pound novelty bot ROBOSCOUT for $699.95, which they advertised as a miniservant that delivers snacks and drinks. The mobile robot was equipped with two adjustable arms that could hold up to two pounds each, obstacle-avoidance sensors, a 2.4GHz transmitter to send video and sound to the robot's 3x4 inch screen, and speech synthesis to give preprogrammed responses to user requests. Novelty robots are not extremely functional, but are entertaining and useful for providing designers feedback about user needs and desires, and robot shortcomings.

Child Minders

The development of child-minders or home companion robots draws from child developmental psychology theories about communication, trust, and play. In the late 1990s, NEC utilized these theories to develop R100, now marketed as the Partner Personal Robot (PA PE RO).

This robot looks like a cross between Star Wars R2D2 and Playskool® Weeble people. Researchers have found that its nonintimidating shape and the huge "eyes" that hide its cameras and other sensor and communication tools motivate children to relate to the robot as a live playmate. The target market is parents whose children are alone after school: the parent e-mails the robot with a message like, "Did you do your homework?" When PA

PE RO crosses paths with the child, it delivers the message vocally, making the message friendlier by singing and dancing. NEC's stated objective is "to create a 'Robot Culture' that provides good human–robot interaction," and since the inception of R100 in 1997, NEC has continually worked with user feedback to make improvements in the robot's image and speech recognition, and behavioral evolution.

PA PE RO is only one of a number of remote-presence robots now on the market. Mitsubishi Heavy Industries first demonstrated its home companion robot, WAKAMARU in 2003. The diminutive yellow humanoid moves on wheels rather than legs, and is equipped with similar communication capabilities as other personal service robots. Mitsubishi spent two years before its release marketing the robot as a member of the family that in addition to monitoring the home and relaying messages to the remote user, could access the internet for weather and news, and keep a person company by discussing the days events. In 2005, the LINUX-powered home companion went on sale in Japan for 1.5 million yen (approx. U.S. $14,000).

An army of small yellow humanoid robots was featured in an episode of the popular American television program, *Crossing Jordan*. These were NUVOs, released in 2005 by the Tokyo-based ZMP Inc. for $6000. These home companion robots look like they were assembled from tinker toys, but have sophisticated communication and control features that have motivated some users to purchase them as substitutes for live pets. The remote presence home companion robots can also function as surveillance robots.

Home Security

In 2000, iRobot introduced the iRobot LE^{TM}, a remote-presence robot that could move around an environment and send information about the location to the user through an online interface. The iRobot LE^{TM} is not a humanoid like the home companions mentioned above, and was not designed for entertainment but to solve the mechanical problems of a robot moving around a human environment, providing thorough visual monitoring of the space, and communicating efficiently with a remote user. The LE moves over flat surfaces on six wheels, and is equipped with two additional wheels in front that can drop down to help it maneuver up and down stairs. A pan-tilt camera is positioned on top of a stalk in the center of the robot which lifts to provide a view of objects above table-top level. A sonar scanner atop the camera provides the robot continuous obstacle detection.

Remote presence differs from teleoperation in that the remote user is not controlling the robots actions from a joystick. Brooks has explained that even with just a momentary internet lag, users can overdirect the robot because they are not seeing a simultaneous reaction to the joystick movement. This is a waste of time that can cause frustration and distrust of the machine's capabilities. Remote presence robots are sent instructions from the user, which they follow locally, using their sensing and other capabilities. This provides a much more fluid and satisfying view for the user.

Although it is unclear what more small nonhumanoid robots could do about an intruder or fire beyond calling for help and relaying information, several have been developed during the past few years. In October 2002 for instance, Fujitsu released the MARON-1 at a retail price of $1625 to monitor entrances or check on pets and relay information and video images to a user's mobile phone. In November 2002, Sanyo and Tmsuk (which produces wireless domestic humanoids as well) released BANRYU, a four-legged guard "dragon," selling at $16,400. BANRYU is equipped with motion and smoke detectors, wireless capability, and speech that enable it to relay messages about danger via howls or phone text messages. Fraunhofer IPA markets a version of its Care-O-Bot as a security robot called Secure-O-Bot. In 2005, the Takashimaya Company in Japan announced it would sell an advanced home security robot in its Japanese outlets for 294,000 yen. The ROBORIOR, designed by Tmsuk, is a low-cost alternative to its 2 million yen domestic robot.

Partners

While DATA of the 1990s television series, *Star Trek: TNG,* was an actor pretending to be an android, recent entertainment robots are approaching the look of actors. In Japan, Professor Hiroshi Ishiguro of Osaka University demonstrated an extremely lifelike android called REPLIEE Q2 expo. This is the latest of several androids developed by Ishiguro, including one that closely resembles a 5-year-old girl. REPLIEE is so lifelike that publicity about "her" raised the question of the *Uncanny Valley*—a term that refers to the point at which a robot's resemblance to humans makes the human uncomfortable. Engineers differ as to the validity of this concept. Ishiguro has said that the close approximation to human looks gives the robot a stronger presence than other humanoids. In any case, one of the few things that give REPLIEE away is the pneumatic air canister that keeps it powered. It can move its head in nine directions, has 42 actuators in its upper body for natural movement, and its soft, flexible synthetic skin is embedded with tactile sensors that enable it to respond to touch. The line between entertainment and engineering has been further narrowed by animatronics like those produced by David Hanson, discussed earlier. Hanson and his team are developing android versions of Albert Einstein, and Philip K. Dick, the author of the 1968 android novel, *Do Androids Dream of Electric Sheep*.

Toys

Robotic toys are among the most promising consumer markets for humanoid and other biologically based robots. The UNECE calculated that in 2005 there were almost 700,000 entertainment robots in use worldwide. Developing robotic toys is attractive to developers because the toy market has a very large existing consumer base, eager to purchase the latest and most impressive gadgets for themselves and their children. Sony's release of a second-generation AIBO (Japanese for friend) in 2000 accompanied the debut of a number of other robotic pets including dogs iCybie by Silverlit of Hong Kong, Dog.com by Tomy, WEB WEB by Lansey (United States), Poo-Chi by Sega, and cats BN-1 from Bandei, ANIM from Tiger, and NeCoRo from Omron. In 2004 ROBOSAPIEN, an interactive, programmable humanoid by veteran NASA engineer Mark Tilden went on the market along with a nonprogrammable battery-operated toy version.

Observing that the popularity of robots like AIBO depend on their ability to keep their owners interested, Frédéric Kaplan, an engineer who designs robotic toys in France, has been creating robotic dogs with a kind of artificial curiosity. These pets are displaying creative, clan-like behaviors that consumers will find amazing and engaging. This activity is in part influenced by the immense popularity of AIBO. Originally developed in 1997, the first run of AIBO pet dogs hit the market in 1999 and sold out quickly despite a price tag of $2,500. AIBO learns to interact with its operators through external stimulation. It can make decisions based on its own judgment from that experience, and displays emotional cues, based on the level of physical and vocal attention paid to them. It displays individuality, moving between dog and puppy personalities at whim. By 2002 AIBO enthusiasts had already figured out how to hack into AIBO and reprogram it, so Sony developed the new software to compete with its own users. It even offers aftermarket hacking software for older models. (Hacking robot software is so popular that iRobot inserted a serial command interface (SCI) into newer models of its Roomba robotic vacuum, and published the SCI specs on its Web site.) Selling for around $1500 in 2005, AIBO continued to be very popular despite the availability of robotic pets like iCybie, which sold for less than one hundred dollars. Some have theorized that the popularity of programmable robotic pets mirrors the activity of training real puppies. This is reflected in an AIBO owner's Web site, which brings together AIBO owners from around the world who refer to their robo-pets with affection and in familial terms.

Researchers have concluded from behavioral studies that human beings relate as powerfully to inanimate objects as to biological ones, particularly when they encounter them in familiar social situation. This may have inspired the development of robotic baby dolls like MY REAL BABY, an interactive, animated doll that could provide realistic emotional responses to a child's interaction with it. According to Hasbro, which teamed with iRobot in 1999 to produce the doll, its Natural Response Technology^{TM} allows the doll to grow emotionally, initiating its own preferences and behaviors. Built on subsumption architecture, the dolls inexpensive sensors and actuator motors provide a low-cost toy at a fraction of the retail price of an intelligent android. In February 2000 MY REAL BABY was introduced at toy trade fairs, and by the end of that year it was on the market, retailing at just under one hundred dollars.

Competing for the market was MGA Entertainment Corporation's MY DREAM BABY, introduced in August, 2000 as a line of unique characters with different ethnic traits. It was engineered with a telescoping spine that allowed it to "grow over time" from a crawling infant to a walking toddler. Voice recognition technology designed by Sensory Inc. of Sunnyvale, California, allows its vocabulary to grow from cooing to hundreds of words learned from interaction with the child. MGA incorporated emotional modeling so that it would respond to the name the child gives it, as well as to hugging, feeding, and other physical human interaction. Publicity for MY DREAM BABY boasted that the doll could revert to infancy so that the child could experience the fun of "being a mommy over an over again" (MGA Press Release).

Mattel released MIRACLE MOVES BABY appeared around the same time. The company boasted that in addition to many of the same attributes as its competitors, the doll had patented flesh-like Flexsoft^{TM} skin, which passed the usual tests for use by children, including fire retardance, resistance to being bitten or chewed, and resistance to ultraviolet sunrays. Designer Caleb Chung called his design "the next iteration of our attempt to recreate life" (Davis, 2000, 271). The overall success of the robot babies is debatable, however, researchers are still producing robot prototypes designed to be attractive to young children. In 2005, a hobbit-like robot designed to interact socially with children was demonstrated at the Robot engineering Center of the Harbin Institute in the northeastern section of China's Heilongjiang Province. This was the first robot produced in China with the capability to make facial expressions. A research group at Yale led by Hideki Kozima uses a robotic infant, INFANOID to explore the use of a robot to assist normal and autistic children in forming social interaction skills. They have developed a contingency-detection game in which the robot reacts to social clues that the child makes and displays those that will induce some response from the child, possibly forming social interaction.

Competitions

In 1989, inventor Dean Kamen founded For Inspiration and Recognition of Science and Technology (FIRST), an organization that teams up professional engineers with students from high schools and technical schools across the country in robot competitions. The idea is to motivate more students to enroll in college science and engineering programs, and thus expand the skill base for robotics. The competition has become so popular that when a national FIRST competition was held at the Disney Epcot Center, it set a new record as the largest non-Disney event ever held there. There is now a FIRST Junior Robotics Program that draws competitors from grammar and junior high schools.

In 1993, robotics engineer Hiroaki Kitano proposed an annual international robot competition as a method of stimulating further research and developing a more collective strategy toward robotics. Kitano's idea was made tangible in the annual RoboCup soccer tournament that has drawn an increasing number of participants since the first event held in Nagoya, Japan, in 1997. Kitano chose soccer as the paradigm because a mobile robot's need to interact with other robots (or people) in a crowded, time sensitive situation requiring attention, dexterity, and decision-making is comparable to a soccer players' need to recognize and follow team members, opponents, and the ball, and to make split-second decisions while maintaining his/her equilibrium. Since soccer is now one of the most popular sports worldwide, it seemed a logical choice to attract a large number of participants. Kitano's goal is for a humanoid robot team to beat the reigning human soccer World Cup holders by 2050. The first soccer bots could do little more than move around the playing field and track the ball. However, by 2001 they were incorporating team strategies into their play. At the 2004 RoboCup in Osaka, Japan, a crowd of 135,000 watched the new humanoid league track the ball, manipulate it using their hands, and pass it with their feet while avoiding obstacles.

RoboCup 2006 Humanoid League competition, Bremen, Germany. Courtesy Sven Behnke

The Eurobot Competition for students aged 18–30 originated in 1994 on a televised French program, *Coupe de France* created by the Planète Sciences to advance scientific knowledge. It has drawn teams from many countries since it became a continent wide competition in 1998. This was also the year that Lego® Mindstorms, programmable robots, first appeared in the United States. The robot kits, which incorporate traditional Lego blocks with actuators and sensor components, are the product of a consortium between the Lego Corporation and MIT. They enable children to build a robot that will do what they want it to do instead of limiting them to preordained functions. Targeted to 9–16 year-olds, Mindstorms has also proved very popular with adults, and is the base robot of choice for many robot competitors. Other high-tech companies like Intel have been sponsoring robot competitions, both to expand their own public profile, and to encourage a new generation to join the field.

Government agencies have been sponsoring robot competitions to draw from a wider technical and innovative base to solve engineering problems. In 2005, Yoseph Bar-Cohen of NASA's JPL organized an Arm Wrestling Grand Challenge to develop more effective robot muscles than are currently

STANLEY crosses the finish line in Primm, Nevada, October 8, winning the 2005 DARPA Grand Challenge. © David Orenstein. Used by permission.

available for prosthetics and robotics. Three teams from the United States and Switzerland competed in the challenge to produce an effective muscle using low-weight electroactive polymers (EAP). The material was chosen because of the comparatively intense force it could provide if the unit is properly designed. All three entries failed against a young student, although research continues in improving artificial muscles.

One of the most widely publicized government-sponsored competitions was the DARPA Grand Challenge. The agency offered a $2 million prize in 2004 to the team that could demonstrate an autonomous land vehicle (ALV) that could traverse a course of 132 miles of mixed desert terrain and thousands of obstacles with no direct communication between the vehicle and the team during the run. It was a considerable engineering feat for any vehicles to finish the race at all. The ALVs had to navigate up and down hills, negotiate sharp curves, maneuver around brush and debris, and drive smoothly near cliffs, just as a military vehicle in a combat zone would have to do. They had to rely only on their mechanics and onboard sensors and computing to perform real-time perception, navigation, and problem-solving. The main challenges are obstacle avoidance without an onboard human driver and reliable vision at high speed. In addition to the perception and computing challenges, team members had to deal with the usual sort of last-minute difficulties encountered by racecar drivers. For example, the Cornell University team had to replace a generator that had seized up the day before the race.

DARPA extended the challenge for a second year when no contenders were able to successfully complete the course. In the 2005 competition, all but two of the 23 ALVs that started went further than the best-performing entry in the 2004 competition, which traveled only 7 miles. Four of the 2005 ALV entries completed the course inside the allotted time; five vehicles finished the course. Two vehicles from CMU under the supervision of William "Red" Whittaker, SANDSTORM and H1GHLANDER, finished second and third, respectively. KAT-5, built by a Louisiana team sponsored by the Gray Insurance Company finished fourth and TERRA MAX, a modified 8-ton military truck finished fifth. STANLEY, the Volkswagon-based ALV built under the supervision of Sebastian Thrun of Stanford AI Lab, finished the track first, winning the $2 million prize.

If engineers have not yet built the perfect robot, they have produced autonomous and semiautonomous machines that assist or substitute for humans in every area of our lives. They have especially made great strides during the last 15 years. The period of time between innovations has narrowed considerably. Reports of improvements in some aspect of robotics, or some new robotic toy appear only months apart. This is in part due to the perception of urgency to find robotic solutions to the dangers of urban and desert warfare, the high cost of health care, and the need for alternative care solutions for a rising geriatric population. It is also due in some way to the desire to fulfill a vision that was born in the archaic imagination. Military, surveillance, and guard robots recall the story of the bronze Talos guarding the Island of Crete. When we encounter personal assistant robots we are reminded of Hephaestos' golden tripods, and the golden servants "with intelligence, speech, and strength."

Conclusion

> ... You don't remember a world without robots. There was a time when humanity faced the universe alone and without friends. Now he has creatures to help him; stronger creatures than himself; more faithful, more useful, and absolutely devoted to him.
>
> —Susan Calvin, Robo-psychologist, in Isaac Asimov's "Robbie"

We cannot say whether the robot was conceived first in the literary imagination or the technological mind. We do know that the idea is rooted in antiquity, and that the initiative to build them has sustained itself through the millennia. Today robots do what our ancestors dreamed they would do, standing in for human workers doing dangerous, tedious, or backbreaking work. While the ancient Egyptians imagined that ushebti would rise and work in the fields of the afterlife for their deceased masters, twenty-first-century engineers have built real robots to do agricultural and domestic work for us. The centuries-old practice of making automata to entertain and to sell ideas and products also remains a key function of robotics, except that today's entertainment robots can be programmed by engineers and hobbyists alike to function in more sophisticated ways than the automata of the past. The integration of robots into so many aspects of society is truly remarkable. Since the 1960s robots have been producing, assembling, or packing many of the products we use; they gather information for us from locales as remote as Mars and as dangerous as compromised nuclear facilities and combat zones. Some of us have already been the beneficiaries of minimally invasive robotic surgery. It is not surprising that many writers and engineers now refer to the current period as the Age of Robots.

Critics and engineers alike have speculated about what this means. In 1982, Ichiro Kato, then Dean of the Graduate School of Science and Engineering at Waseda University who led the first humanoid project, predicted that by 2000 robots would be so prevalent in our lives that they would come to be referred to as "My Robot." Furthermore he wrote that though bipedal robots were unnecessary for the time he was writing, using them to study human locomotion would have practical uses by the twenty-first century. Kato was right on both counts. As we have seen many engineers argue that the work they have done in bipedal locomotion will indeed be useful beyond entertainment robots.

Hans Moravec, a graduate student when we met him in Chapter 5, is now head of the mobile robotics lab at Carnegie Mellon. Almost 20 years ago he began speculating that by around 2020 we will have developed universal robots with the processing capabilities of human beings. According to Moravec, they will be able to simultaneously simulate (imagine) a world and reason about that simulation. They will have the physical and intellectual skill to anticipate outcomes, to respond to emotional cues, and to communicate with each other in sophisticated ways, to avoid physical obstacles, and respond to emergencies. As we have seen, robots can certainly *act* intelligent and emotional today, and their many mechanical skills make it possible for them to function in useful, specified ways. They may not meet Moravec's timeframe for self-awareness (even he has recently been quoted as having adjusted the timeframe for fourth-generation humanoids to around 2040). However, many engineers have indeed committed to instilling robots with emotional and intuitive skills.

A few years ago Rodney Brooks identified two capabilities robots need for this to truly be the "Age of Robots": a vision system that would enable a robot to recognize an object like a cell phone, and the mechanical ability to pick it up and correctly make a call (David Whitford, *Fortune*, May 2, 2003). These actions would implicitly involve some understanding of the point of doing the task. In the past few years, however, engineers have made inroads into dexterity and haptics, AI, and robot stability that have, for instance, improved robots' ability to sense things like force, temperature, and light. They have improved balance to the point that machines can leave the ground for short periods and land stably. If they cannot understand the process of making a phone call yet, many have the mechanical ability to manipulate the buttons. Researchers have solved some of the problems of navigation so that vehicles can operate autonomously on land, and in the air.

Joseph Engelberger considered the development of humanoid robots as a kind of craze, arguing that these robots don't do anything useful. Japanese robotics engineers who are working on humanoids think that entertainment is first something important, and second, only a step in a future direction. The severest critics of robotics seem to be unwilling to put the development of the field in historical perspective. To say for instance, that humanoids like ASIMO or QRIO can "only" run or jump is to ignore the fact that only a few decades ago, they could not even walk independently. They seem to miss the achievement of many electrical and mechanical engineers to coordinate the weight, position, and actuation of limbs, the control and feedback for this to happen. Critics perhaps misunderstand the marketing of intelligent robot prototypes for entertainment purposes as a final, trivial aim; while researchers describe robotic toys and trade show mascots as test beds for human–robot interaction and socialization. Improvements in speech and facial recognition and synthesis, balance, and bipedal locomotion, and sensors being built into the next generation of consumer robots like ROBOSAPIEN and ROBORAPTOR will arguably be valuable across the industry. As I pointed out in the last chapter, vision and navigation systems developed by companies like Evolution Robotics are indeed being used in both entertainment and service products.

Provocative titles of magazine articles like "A Robot Revolution is Coming Your Way," "The Robots Are Here," and "Human Being 2.0: The Race to Make Androids that Walk, Talk, and Feel Just Like the Rest of Us," create public anticipation for the imminent appearance of androids like Star Trek TNG's DATA and David of the film AI. Despite the fact that authors of articles like this disclaim that androids are "right around the corner," their illustration-packed articles play down the problems explained in the text, like the still short operating time between battery charges, and differentiating one person from another.

Still, the response to robots and androids in the home is mixed. A survey conducted by BBC News, inspired by the UNECE projection of a sharp increase in domestic and entertainment robots by 2007, asked people to comment on the idea of having a robot do chores in their home. A contributor from Scotland thought he would welcome a domestic robot that could do chores as long as it was not too expensive and properly conditioned to work in the home. Likewise, a reader form Taiwan looked forward to a robot that could, for instance, help her baby if it fell out of bed, provided the unit was affordable; while an AIBO owner from Germany was against the idea of a robot babysitter. More than one reader found the idea "scary," while many reminded fellow readers that robots are after all, "just machines." One reminded readers how many people there were looking for the kinds of jobs domestic robots would do, while another thought we should be developing robots to help out in underdeveloped countries instead of serving us in our homes. Polls like this one may seem trivial, but they are keeping the discourse of the robot active.

Some people question the direction service robotics is taking. Will we realize the goals of Japanese engineers who are trying to build friendly robotic helpers; or will we build a new generation of war machines and leave the sick and elderly to an already overtaxed health care system? Does one achievement necessarily negate the other? Will efforts to explore other planets and perhaps create settlements be stifled by debate over whether humans or robots should do that work? Will new innovations in AI and robotics push too high a percentage of the already small manufacturing workforce out of jobs, or make room for new kinds of human occupations?

Mark Tilden, developer of minimally programmed, often solar-powered biorobotics, and inventor of ROBOSAPIEN for the Korea-based toy company, WowWee, said in a 2000 interview that he hoped that the robots that were being developed for war would soon be adapted for peaceful, life-saving applications. His concern may be related to the source of funding for robots, since in the United States, the majority of R&D money comes from government agencies like DARPA for developing robots for military use. In Europe, the tendency has been to study AI learning and A-life, while the Japanese have created a long-term initiative to develop robots as helpmates in many areas of life, from civil service to business, to the home. While military projects yield innovations that would be useful in other areas of life, the concern is whether such applications will be made.

The future of robots depends on continual innovation, and that is why the initiatives to attract grammar and high-school students to the computing and engineering sciences through funded robot competitions are vital. We need not all be engineers, however, to understand that as with any technology, robots will have the most value where people can grasp their advantage as tools, and not as toys. They will be a benefit if we consider and take responsibility for the social, economic, and political implications of their widespread use. We cannot afford to be only consumers of established technologies. We must contribute to a wider discussion about how a technology is developed, distributed, and used, and whether its use impinges on others in a positive or negative way. Fictional visions of robots running amok are only as realistic as the willingness of society to opt out of the decision-making process when the time comes to fund and make policy on new technologies.

Robots, like any other machine, are considered advantageous when the useful work they can do equals or exceeds the effort (power, expense) to manufacture, distribute, and operate them. This is one reason why the percentage of humanoid robot labs is far smaller at the moment than for simpler mobile robots like robotic vacuums or pool cleaners, or small robotic toys. They are relatively simple to operate and control, efficient, and affordable to the average middle-class consumer.

Only the technological facet of the life of the robot has been addressed in depth here. The limitation of space could not allow me to fairly deal with the reception of robots by organized labor, and the economic impact on workers and businesses. I could make only a passing reference to the response of consumers to robots and to the influence of the human imagination on the direction this technology has taken thus far. Two other important components are the story of career paths of the individuals whose inventions are touching our lives, and the institutions that have made their research possible. The life story of the robot cannot be fully understood without connecting it to the story of the humans who imagined, built, and use them; but those stories would fill volumes. I encourage those with a serious interest in studying the history of robots to explore these connections. A good starting point is the *Further Reading* section that follows. There, in addition to a list of books used in the writing of this volume, are Web sites of robotics labs, a selection of engineering reports, and a list of robotics organizations.

Glossary of Terms

Actuator. A mechanism that translates power into motion, e.g., electric motors, hydraulic cylinders, or rotary actuators.

Android. A humanoid robot with close resemblance to the physical features and behaviors of a human being.

Anthrobot. A robot with the general looks and functions of a human.

Anthropomorphic. Having a human-like form.

Arm Geometry. Refers to the directions and angles a robotic arm can move within its workspace, e.g., Cartesian (geometric), Polar, Revolute, or Cylindrical.

Artificial Intelligence (AI). The demonstration of decision-making, problem-solving, analytical thinking by machines using symbolic or environmental data. AI refers to computer software, but is also used to describe a robot equipped with AI.

Automation. The processing of materials by devices such as robots that can make and execute decisions with little or no human intervention through the use of self-correcting control systems.

Automaton. A machine that has "the form of an organized being [containing] within itself a mechanism capable of creating movement and simulating life." (*Encyclopédie du XIXème siècle*, Paris, 1877).

Behavior Based Robotics (BBR). BBR are minimally programmed to navigate environments and learn from their experience via feedback, thus expanding their capabilities. Typically their programming is layered, from simple to complex. Also called "stimulus-response" robots.

Biomechanics. The study of living things as mechanical structures.

Biomimetic. Mimicking natural biology.

Bipedal/Bipedalism. "Two-footed."/The ability to walk stably on two legs.

Cam. A rotating disc with outer edges of varying shapes, used to program machines so that they can repeat a sequence of motions. The more complex the desired movement, the more cams are required.

Cartesian coordinate geometry. A mode of robot arm movement based on Cartesian system for graphing mathematical functions. All axes of movement are perpendicular to each other.

Computer-aided Manufacturing. The use of specialized computers to control, monitor, and adjust tools and machinery in the manufacturing process.

Connectionism. An approach to studying intelligence. Storing and using problem-solving information or knowledge as a pattern of connections between a large number of simpler processing units that operate in parallel. Neural net technology is a result of this approach.

Control engineering. The study and invention of methods of regulating machine behaviors such as speed, force, temperature, and direction.

Cybernetics. A control theory based on the idea that an animal's sensorial experience with its environment is communicated as feedback to its central nervous system, which in turn regulates systems like temperature, fright, fatigue, etc. A method of regulating machines such as thermostats, automatic weapons, and robots.

Degrees of Freedom (DOF). The number of independent directions of motion a robot or one of its effectors can make, described in angles. In a serial robot each joint represents one DOF. A three degrees of freedom (3DOF) system moves along the X (horizontal), Y (vertical), and Z (depth) axes. A 6DOF system moves along X, Y, and Z as well as pitch (up and down, like a box lid), yaw (left and right, like a door on hinges), and roll (clockwise or counterclockwise rotation).

Dynamic Balance. The ability of a robot to balance while moving despite changes in the terrain. In two-legged robots, balance while one foot is off the ground. (See ZMP)

Endeffector. A tool attached to the robot's wrist plate, such as a paint sprayer, a gripper, or an artificial hand.

Feedback. A signal such as an error message sent from an active circuit or a device back to the input. One or more sensors and a method of communicating sensor readings to the onboard or remote processor enables a robot to adjust operating elements such as speed, temperature, direction, or pressure.

Flywheel. A disc that regulates machine speed and uniformity of motion.

Friction. Force resisting motion between two objects that are in contact with each other.

Haptic sense. The ability to feel such things as temperature, force, or texture. Haptics research has become an important part of robotics engineering because it is expanding the versatility and precision of industrial and service robots.

Inertia. The tendency of a body to remain either in motion or at rest unless acted upon by another force.

Integrated circuit. Invented by Jack St. Clair Kilby at Texas Instruments in 1958. An electronic circuit built on a semiconductor matrix like silicon.

Inverse kinematics. The process in which a computer calculates the joint angles necessary to change the orientation of the robot's endeffector.

Kinematics. —Kinematics is the science of motion. It is restricted to a pure geometrical description of motion with reference to position, orientation, velocity, and acceleration.

Localization. The process by which a robot works out where it is.

Manipulator. Another term for a robotic machine. The arm, shoulder, and endeffector comprise a machine that manipulates objects in different ways, e.g., lifting, packing, boring, and welding.

Master/slave control. An electronic interaction in which one device acts as the controller (the master) and initiates the commands, and the other devices (the slaves) respond accordingly.

Mechatronics. The integration of mechanics, electronics, computer, and control disciplines. Together they result in the generation of more intelligent, flexible, versatile, reliable, and simple systems.

Mind/Body Problem. A philosophical question: how does the "mind" (thought), which is nonphysical, emerge from the physical brain? This question is the starting point for the debate between those who believe that machines can be intelligent, and perhaps even conscious, and those who do not.

Moore's Law. In the 1960s, Gordon Moore, the former CEO of the Intel Corporation, correctly predicted that the size of each transistor on an integrated circuit chip would be reduced by 50 percent every 18 to 24 months, doubling the number of components and speed per chip, thus doubling computing power.

Neural Network. A computer simulation of human neurons (the information-processing cells of the central nervous system) that is meant to emulate the computing structure of human brain neurons. More flexible than traditional programming methods.

Payload. In the case of a robot, the weight capacity of the structure. A robot whose back and legs have a payload of 100 grams cannot handle an onboard processor that exceeds that weight.

Pitch. Up and down motion of a robot limb.

Prime Mover. The main power source for a machine's motor, e.g., water, wind, electricity.

Rangefinder. An active sensor used to find the distance of objects in the environment.

Revolute. Rotary

Robustness. The ability of a system with a fixed structure to perform any required multiple functional tasks in a changing environment.

Rocker bogie. "Bogie" was originally used to refer to a train undercarriage with six wheels that can swivel to curve along a track; while "rocker" is a differential design that keeps the chassis balanced by "rocking" up or down depending on the various positions of the wheels. "Rocker-bogie" carriage systems are used in mobile robots like the Mars rover SOJOURNER.

Roll. Rotation of a joint in the axis that is in line with the arm.

Sensor. A device that responds to physical stimuli like heat, light, sound, pressure, magnetism, motion. It transmits the resulting signal as feedback, which is then acted upon by a servo or other control mechanism. (e.g., photocell, thermometer)

Servomechanism. A system used in complex machines that automatically controls functions like velocity, force, position, or acceleration. A feedback system forces the servomotor to correct any error (difference) between the intended or commanded output and the actual output.

Singularity. A term commonly used to indicate a position of the robot where a particular mathematical formulation fails.

Subsumption architecture. A BBR paradigm developed by Rodney Brooks at MIT involving a layered approach to programming and learning. (See BBR above.)

Teach Pendant. A portable device used to control robot movement. Positional data points are generated and stored by recording the movement of the robot arm through a path of intended motions within a determined space.

Teleoperation. The remote control of a robot or other device.

Thyrotron. A switching device that uses one of several gases like Mercury or Hydrogen.

Transistor. An electronic device made of semiconducting material used as an amplifier, rectifier, detector, or switch.

Uncanny Valley. A term coined by the Japanese engineer Masahiro Mori to identify the point at which people would cease to feel familiar and positive about robots and begin to feel unease. He located the "Valley" or emotional plunge in acceptance at the point where robots become too lifelike.

Wetware. The name that computer scientists and engineers give to the human brain and nervous system, to distinguish them from computer hardware and software.

Yaw. Side-to-side motion of a robot joint.

Zero Moment Point (ZMP). The point where the combined forces of gravity and inertia working on the robot intersect with the ground. Important to the stability of walking robots.

Further Reading

BOOKS

Altick, Richard D. *The Shows of London*. Cambridge, MA: Harvard University Press-Belknap, 1978.

Asimov, Isaac. *I, Robot*. New York: Doubleday-Spectra/Bantam, 1991.

Bailly, Christian. *Automata: The Golden Age, 1848–1914*. London: Robert Hale, 2003 [1987].

Bennett, S. *A History of Control Engineering 1930–1955*. London: Peter Peregrinus for IEE, 1993.

———. *A History of Control Engineering 1800–1930*. London: Peter Peregrinus for IEE, 1979.

Billingsley, J., ed. *Robots and Automated Manufacture*. IEEE Control Engineering Series Vol. 28. London: Peregrinus/IEEE, 1985.

Brooks, Rodney A. *Flesh and Machines: How Robots Will Change Us*. New York: Pantheon, 2002.

Cardwell, Donald. *Wheels, Clocks, and Rockets: A History of Technology*. New York: Norton, 1995.

Carroll, Dr. Terrence, V.P. for Support Services, Magee Women's Hospital, Pittsburgh, PA, House Committee on Science, Space, and Technology, Subcommittee on Science, Research and Technology, *Robotics Technology and Its Varied Uses*, 101st Cong., 1st sess., 1989, Committee Print 56.

Chapuis, Alfred and Edmond Droz. *Automata: A Historical ad Technological Study*. Trans. Alec Reid. Neuchâtel: Editions du Griffon, 1958.

De Vaucanson, Jacques. *Le Méchanisme du Fluteur Automate (An Account of the Mechanism of an Automaton or Image Playing on the German Flute.)* Paris 1738, Trans. J.T. Desaguliers, London: 1742 and rpt. In Series I, no. 5 The Flute Library ed. by Frans Vester, preface by David Lasocki, The Netherlands: Frits Knuf-Buren (Gld), 1979.

Franklin, Stan. *Artificial Minds.* Cambridge, MA: MIT, 2001.

Geduld, Harry M., and Ronald Gottesman, eds. *Robots Robots Robots.* Boston: New York Graphic Society, 1978.

Gimpel, Jean. *The Medieval Machine: The Industrial Revolution of the Middle Ages.* New York: Penguin, 1976.

Hillier, Mary. *Automata & Mechanical Toys.* London: Jupiter, 1976.

Ichbiah, Daniel. *Robots: From Science Fiction to Technological Revolution.* Trans. Ken Kincaid. New York: Abrams, 2005.

Iovine, John. *Robots, Androids, and Animatrons: 12 Incredible Projects You Can Build.* 2nd. ed. New York: McGraw-Hill, 2002.

Kurzweil, Ray. *The Age of Spiritual Machines: When Computers Exceed Human Intelligence.* New York: Penguin, 2000.

Logsdon, Tom. *The Robot Revolution.* New York: Simon and Schuster, 1984.

Woolley, Benjamin. *The Bride of Science: Romance, Reason, and Byron's Daughter.* New York: McGraw-Hill, 1999.

Minsky, Marvin. "A Framework for Representing Knowledge." In *The Psychology of Computer Vision*, Patrick and Winston, eds. New York: McGraw Hill, 1975, 211–279, quoted in Crevier, Daniel. *AI: The Tumultuous History of the Search for Artificial Intelligence.* New York: Basic, 1993, 172–173.

Moravec, Hans. *Robot: Mere Machine to Transcendent Mind.* New York: Oxford University Press, 1999.

Morus, Iwan Rhys. *Frankenstein's Children: Electricity, Exhibition, and Experiment in Early Nineteenth Century London.* Princeton: Princeton University Press, 1998.

Ord-Hume, Arthur W.J.G. *Clockwork Music.* London: Allen & Unwin, 1973.

Payes, Rachel Cosgrove. "Grandmother Was Never like This." In *Androids, Time Machines, and Blue Giraffes*, Roger Elwood and Vic Ghidalia, eds. Chicago: Follett, 1973, 52–55.

Ritchie, David. *The Computer Pioneers.* New York: Simon and Schuster, 1986.

Rosheim, Mark E. *Robot Evolution: The Development of Anthrobotics.* New York: Wiley, 1994.

Schodt, Frederik. *Inside the Robot Kingdom: Japan, Mechatronics, and the Coming Robotopia.* Tokyo: Kodansha, 1988.

Scientific American, eds. *Understanding Artificial Intelligence.* New York: Warner, 2002.

Ure, Andrew. *The Philosophy of Manufactures* (London: 1835), 13, quoted in S. S. Schweber, "Scientists as Intellectuals: The Early Victorians," in *Victorian Science and Victorian Values: Literary Perspectives*, James Paradis and Thomas Postlewait, eds., *Annals of the New York Academy of Sciences* 360 (April 20, 1981): 14 and notes 66, 67.

ARTICLES

Adams, Bryan. "Meso: A Biochemical Subsystem for a Humanoid Robot." Available from http://www.ai.mit.edu/projects/humanoid-robotics-group/Abstracts2000/adams.pdf.

Chen, B. and L.L. Hoberock. "Machine Vision Recognition of Fuzzy Objects Using a New Fuzzy Neural Network." *Proceedings of the International Conference on Robotics and Automation* (IEEE/ICRA) (1996).

Davis, Eric. "Congratulations, It's a Bot!" *Wired* (Sept. 2000): 266–271.

Dreyfus, Hubert. "Why Symbolic AI Failed: The Commonsense Knowledge Problem." Lecture presented at UH, January 27, 1998. Available from http://www.hfac.uh.edu/phil/garson/DreyfusLecture1.htm.

Droz, Edmond. "From Jointed Doll to Talking Robot." *New Scientist* 14(282) (1962): 37–40.

Halme, Aarne, Peter Jakubik, Torsten Schönberg, and Mika Vainio. "Controlling the Operation of a Robot Society through Distributed Environment Sensing." *Human-Oriented Design of Advanced Robotics Systems (DARS).* Ed. P. Kopacek. IFAC Workshop held in Vienna, Austria, September 1995.

Hashimoto, S., et al. "Humanoid Robots in Waseda University–Hadaly-2 and Wabian." *Autonomous Robots* 12 (2002): 25–38.

Kosko, Bart. "Fuzzy Logic." *Understanding Artificial Intelligence.* New York: Warner, 2002: 33–36.

Kozima, Hideki. "The Infanoid Demonstration." *Second Workshop On Robotic And Virtual Interactive Systems In Therapy of Autism and Other Psychopathological Disorders*, Paris, September 27, 2002. Available from http://rescif.risc.cnrs.fr/seminaires/archives/2002-2003/robot_autism.html.

Kulyukin, Vladimir. "Human-Robot Interaction through Gesture-Free Spoken Dialogue." *Autonomous Robots* 16 (2004): 239–257.

Lenat, Douglas B. "From 2001 to 2001." Available from: http://www.cyc.com/halslegacy.html.

McCarthy, J., M.L. Minsky, N. Rochester, and C.E. Shannon, "A Proposal For the Dartmouth Summer Research Project On Artificial Intelligence," August 31, 1955. Available from http://www.formal.staford.edu/jmc/history/dartmouth/node1.html.

Metta, Giorgio and Paul Fitzpatrick. "Development of Imitation in a Humanoid Robot," Cambridge, MA: MIT AI Lab. Available from http://www.ai.mit.edu.

Rao, Rajesh P.N. and Andrew Meltzoff, "Imitation Learning in Infants and Robots: Towards Probabilistic Computational Models." *Proceedings of Artificial Intelligence and Simulation of Behavior (AISB) 2003: Cognition in Machines and Animals.* Available from www.cs.washington.edu/homes/rao/Rao_Meltzoff_AISB03.pdf.

Scassellati, Brian. "Theory of Mind for a Humanoid Robot." MIT AI Lab. Available from http://www.ai.mit.edu/projects/humanoid-robotics-group/Abstracts2000/scaz.pdf.

Schaal, Stefan, "Is Imitation Learning the Route to Humanoid Robots?" *Trends in Cognitive Sciences*, 3 (1999): 233–242.

Simon, Herbert and Allen Newell. "Heuristic Problem Solving: The Next Advance in Operations Research," *Operations Research*, 6 (1957): 1–10.

Swinson, Mark L. and David J. Bruemmer. "Expanding Frontiers of Humanoid Robotics." In *IEEE Intelligent Systems Special Issue on Humanoid Robotics* (July/August 2000): 9 pp. Available from: http://www.inel.gov/adaptiverobotics/humanoid robotics/publications/special-issue.pdf.

Takanishi, Atsuo, Sang Ho Hyon, Samuel Agus Setiawan and Jin'ichi Yamaguchi, "Physical Interaction Between a Human and Humanoid Through Hand Contact." *Advanced Robotics* 13(3) (1999): 303–305.

Taylor, Geoffrey and Lyndsay Kleeman. "Grasping Unknown Objects with a Humanoid *Robot. Proceedings of the 2002 Australian Conference on Robotics and Automation*, November 27–29, 2002: 191–196. Available from http://www.ecse.monash.edu.au/centres/irrc/LKPubs/Taylor-Kleeman.pdf.

Turing, Alan. "On Computable Numbers, with an Application to the Etscheidungsproblem." Proceedings of the London Mathematical Society second series (November–December 1936): 230–265. Available from at http://www.abelard.org/turpap/turpap.htm.

———. "Computing Machinery and Intelligence." Mind (October 1950): 433–460.

White, Lynn Jr. Cultural Climates and Technological Advance in the Middle Ages." *Viator* 2 (1971): 171–201.

Ye, D., H. Mazafarrari-Naeini, C. Busart, and N.V. Thakor. "MEMSurgery: An Integrated Test-bed for Vascular Surgery. *Robotic Publications Ltd.* (September 20, 2004): 21–30. Available (fee) from: http://www.roboticspublications.com.

Zimmer, Carl. "How the Mind Reads Other Minds." *Science* (May 16, 2003): 1079–1080.

RESEARCH LABS AND ROBOTICS COMPANIES WEB SITES

ABB Robotics: http://www.abb.com/robotics

Boston Dynamics: http://www.bostondynamics.com/

Carnegie Mellon (CMU)
    – Computer Science department: http://www.cs.cmu.edu/
    – CMU Robotics Institute: http://www.ri.cmu.edu/

Case Western Reserve University Biorobotics Lab: http://biorobots.cwru.edu/

Evolution Robotics: http://www.evolution.com/

Gecko Systems Service Robots: http://www.geckosystems.com/

Fraunhofer Institute of Autonomous Intelligent Systems: http://www.ais.fraunhofer.de/ROBLAB/index.en.html

Fraunhofer IPA Care-O Bot robots: http://www.care-o-bot.de/english/Care-O-bot_2.php

Fujitsu HOAP series robots: http://www.fujitsu.com/global/about/rd/200506hoap-series.html

Hanson Robotics (K-BOT and other androids): www.hansonrobotics.com

Honda ASIMO technology: http://asimo.honda.com/inside_asimo.asp?bhcp=1

Idaho National Engineering and Environmental Laboratory (INEEL) Adaptive Robotics: http://www.inl.gov/adaptiverobotics/humanoidrobotics/

Intuitive Surgical (DAVINCI® surgical system): www.intuitivesurgical.com

iRobot: http://www.irobot.com/

Johns Hopkins University Haptic Interaction Lab: http://www.haptics.me.jhu.edu/

MIT Artificial Intelligence Lab (CSAIL): www.csail.mit.edu/

MIT Humanoid Robotics Group: http://www.ai.mit.edu/projects/humanoid-robotics-group/
    – COG: http://www.ai.mit.edu/projects/humanoid-robotics-group/cog/cog.html
    – MIT KISMET: Sociable Machines - http://www.ai.mit.edu/projects/sociable/kismet.html

KISMET videos: http://www.ai.mit.edu/projects/sociable/videos.html
    – Leg Lab: http://www.ai.mit.edu/projects/leglab/

Mitsubishi Heavy Industries
    – WAKAMARU domestic robot: http://www.mhi.co.jp/kobe/wakamaru/english/

NASA ROBONAUT: http://robonaut.jsc.nasa.gov/
    – ROBONAUT videos: http://robonaut.jsc.nasa.gov/videos.htm
    – JPL Lab robotics: http://www-robotics.jpl.nasa.gov/

PaPeRo Personal Robot: http://www.incx.nec.co.jp/robot/english/robotcenter_e.html

Toyota PARTNER Robots: http://www.toyota.co.jp/en/special/robot/

RedZone Robotics: www.redzone.com

Shadow Robot: www.shadowrobot.com

Stanford Artificial Intelligence Lab (+robotics): http://ai.stanford.edu/

Stanford Research International (SRI): http://www.sri.com/

TMSUK Robotics: http://www.tmsuk.co.jp/english/

University of S. Florida Center for Robot Assisted Search and Rescue (CRASAR): http://crasar.csee.usf.edu/MainFiles/index.asp

Waseda University Humanoid Robotics Institute: http://www.humanoid.waseda.ac.jp/

WowWee Robotics (makers of ROBOSAPIEN): http://www.wowwee.com/

PROFESSIONAL ASSOCIATIONS

AAAI American Association for Artificial Intelligence: www.aaai.org

Association Espanola de Robotica (Spain): http://www.aeratp.com

BARA British Auto and Robot Association: http://www.bara.org.uk

First Robotics (For Inspiration and Recognition of Science and Technology.) Inspires an appreciation of science and technology: www.usfirst.org

CHONNAM National University (South Korea): http://micronanorobot.re.kr
Croatian Robotics Society (Zagreb) (no Web site available) Tel: +38 5148 48 760
DIRA Danish Industrial Robot Association: http://www.dira.dk
IEEE Institute of Electronics and Electronic Engineers, Inc.: www.ieee.org
IFR International Federation of Robotics: www.ifr.org
IPA (Germany): http://www.ipa.fraunhofer.de
JARA Japanese Robotics Association: http://www.jara.jp
KOMMA Korea Machine Tool Manufacturer's Association: http://www.komma.org
ÖGART Austrian Society for Automation and Robotics: http://www.ihrt.tuwien.ac.at
Precarn Associates, Inc. (Ontario, Canada): email: johnston@precarn.ca
RIA Robotic Industries Association (USA): www.roboticsonline.com
RoboCup robot soccer competition official Web site: www.robocup.org
Robotics Society in Finland: http://www.roboyhd.fi/english/hallitus.html
Schweizerische Gesellschaft für Automatik: http://www.sga-asspa.ch
Singapore Industrial Automation Society: http://www.esiaa.com
SIRI Italian Association of Robotics (no Web site available) Tel: +39 02 262 552 57
SYMOP (France): www.symop.com
SWIRA Swedish Industrial Robots Association: http://www.swira.org/hemsida/default.asp (in Swedish) Contact: Mr Thomas Hardenby Tel: +46-8-783-80-00 Fax: +46-8-660-33-78
TBL Federation of Norwegian Engineering: http://www.tbl.no
VDMA (Frankfurt, Germany): http://www.vdma.org

OTHER RELATED WEB SITES

Android World—www.androidworld.com: devoted to humanoid robotics. Focuses on humanoid and other biologically based robots. Links to research sites of the robots in progress.
Hellbrunn Park, Salzburg, Austria. Images of its trick fountains and automata available from http://www.hellbrunn.at/hellbrunn/english/trickfountains/mechanical_theatre.asp.
Japanese automata: http://int.kateigaho.com/spr05/robots.html
Japanese Mechanical Archer (1850s) reproduced by Shoji Takashina (1998): http://int.kateigaho.com/spr05/robots.html
Leonardo's Robot from BBC News: http://news.bbc.co.uk/1/hi/sci/tech/149724.stm
Maillardet's Writing Automaton at Franklin Institute, and information about automata available from: http://www.fi.edu/pieces/knox/automaton/
Tipoo's (Tipu's) Tiger Automaton at Victoria and Albert Museum: http://www.vam.ac.uk/collections/asia/object_stories/Tippoo's_tiger/index.html

Note: URLs were active and current as of June 2006. This is a selected list and in no way represents all robotics labs or companies.

Index

The letter f following a page number denotes a figure.

Rosheim, Mark: Leonardo's schematics for automata, 17
ROTOPOD, 137, 138
Roullet and Decamps automaton firm, 41, 42
ROV, xxiii, xxiv; Triton XL, xxxi; Vorcity Control Unmanned Undersea Vehicle (VCUUV), 149
Ruina, Andy. See Passive/dynamic walking

Saint-Germain-en-Laye, Château, 17
SANDSTORM. See DARPA, Grand Challenge
Sawada, Hideyuki'. See Synthetic vocal tract
Scassellati, Brian, and Theory of Mind, 132
Scheinman, Victor: founds Automatix, 70, 96; STANFORD ARM, xxv, 6–69; Vicarm, Inc., xxv, 69
Schickard, Wilhelm, and early calculating machine, 60, 61
Schlottheim, Hans, 27
Schockley, William Bradford, and transistor, 76. See also Brattain, W.H., and Bardeen, John
Scientific Management, 47, 48
SCOOBA (iRobot), 152
Screw (water), and Archimedes, 8
SDR (Sony Dream Robot). See Sony
Searle, John, and criticism of GOFAI, 91
Segway®: Human Transporter (HT); Robotic Mobility Platform (RMP), 148
Selfridge, Oliver, and PANDEMONIUM, 78. See also Pattern recognition software
Sensors, 60, 61, 69, 72, 74, 113, 115, 120, 138, 142, 143, 150, 152, 154; in ALVs, 159; Haptic sensors, 127; improvements for industrial robots, 123; in robot dolls, 156

Servomechanisms, 55
Shabti, Egyptian, xii, 4, 5. See also Ushebti
Shadow Robot Company, xxviii, 118; dexterous hand, xxxiii, 128
SHAKEY, xxiv, 100–102
Shannon, Claude, and binary calculator, 63; mathematical model of communication, 76
Shaw, Clifford J.: LOGIC THEORIST, 78. See also Newell, Allen; Simon, Herbert
Shortliffe, Edward. See MYCIN
SHRDLU. See Winograd, Terry. See also Artificial intelligence
Silicon Valley, 76
SILVER ARM (David Silver), xxv, 69. See also Robots, industrial, small parts handling
Simon, Herbert, prediction about computer intelligence, xxiv; Carnegie Mellon, 78; and LOGIC THEORIST, 78; on Minsky's frames, 89; quotation, 75;
SNARK. See Unmanned Aerial Vehicle (UAV)
SOJOURNER, 143, 144
Sonar, 100, 103, 108, 109
Sony, xi; AIBO, xxx, xxxiv, 155, 156 SDRX, xxxi; QRIO, xxxiii, xxxiv, 134, 135; SDR-4X, xxxii
Speech recognition, DECIPHER, 104
Speech synthesis, 133, 134
Spencer, Christopher, and cam-operated lathe, xx
Spinning jenny. See Hargreaves, James
SPIRIT and OPPORTUNITY, xxxiii, 144, 145
Stanford Research Institute (SRI): CART, 102, 103; Stanford Artificial Intelligence Lab (SAIL), 100, 102. See also Baumgart, Bruce; Moravec, Hans; ORM; Scheinman, Victor; SHAKEY; SILVER ARM

About the Author

LISA NOCKS is a historian who writes on the diffusion of scientific and technical knowledge to the public through the media. She has authored a number of articles on the history of media technologies and essays on the relationship between science fiction and science. She is currently teaching at Fordham University in the department of Communication and Media Studies.